Starman

Starman

the Truth behind the Legend of
Yuri Gagarin

JAMIE DORAN + PIERS BIZONY

BLOOMSBURY

Picture Acknowledgements

Two great rivals: Gherman Titov (left) and Gagarin
(*Baker/Planet*), Loyal friend and fellow cosmonaut
Alexei Leonov (*Baker/Planet*), Suited up prior to
flight, 1961 (*Baker*), A view through the Vostok
capsule's hatch (*Baker*), A Vostok on display showing
the troublesome electrical connector (*Baker/Planet*),
A Vostok ejection seat and space suit (*Baker*), Gagarin
and his political champion, Nikita Khruschev, celebrating
in Moscow, 14 April 1961 (*Baker*), Gagarin with Sergei
Korolev shortly after the flight, 1961 (*Baker/Novosti*),
Konstantin Feoktistov, Boris Yegorov and Vladimir Komarov
after their mission aboard Voskhod I, October 1964 (*Baker/TASS*)

First published in 1998 by
Bloomsbury Publishing plc
38 Soho Square
London W1V 5DF

Copyright © 1998 by Jamie Doran and Piers Bizony

The moral right of the authors has been asserted

A copy of the CIP entry for this book
is available from the British Library

ISBN 0 7475 3688 0

10 9 8 7 6 5 4 3

Typeset by Hewer Text Ltd, Edinburgh
Printed in Great Britain by Clays Ltd, St Ives plc

CONTENTS

Dedicated to the memory of
Olga Alexandrei and Igor Nosov

FOREWORD

The Soviet period of history is, for some, a time best forgotten. The massacres of the Purges, the horrors of the Gulags and forced labour, the fictitious Five Year Plans and the Orwellian lack of personal freedom hardly make for happy memories. The other side of the coin, however, is that many of humankind's greatest achievements came from a nation that had to be almost entirely rebuilt after the devastation of the Second World War. Twenty-seven million Russians died fighting the Nazis, and the entire infrastructure of their nation was disrupted – but after barely a decade, the Soviet Union was a global superpower, with new world-beating technology capable of reaching into space.

Yet, for all but a very few elderly die-hards, there are no great heroes from the Soviet era. Indeed, to be a hero at all under communism was itself an anomaly, for no individual could be greater than the collective whole. No, there were no great heroes . . . except for one.

At a giant crossroads on Moscow's Leninsky Prospekt a 30-metre-high steel statue of him dominates the skyline, while ninety-six kilometres to the north-east another memorial marks the spot where, just a few days after his thirty-fourth birthday, he died in a terrible accident that gouged a huge crater in the frost-hardened earth. Three decades after this catastrophe, fresh bouquets of flowers at the site confirm the continuing love of ordinary citizens for this man. Young or old, privileged or poor, most Russians still smile with pride at the merest mention of his name. They recall with genuine affection a peasant boy with a winning smile, who stunned the entire world with his achievement. Russians don't hesitate to remind us Westerners, 'He was the first, you know.'

1

The first man in space: Yuri Alexeyevich Gagarin.

My fascination with Gagarin was sparked three years ago, when I was completing a major television documentary series on the Soviet nuclear-weapons programme. It became clear to me that the story of *The Red Bomb* would not be complete without a description of the rockets that were built to carry them out into space, even as they threatened us with destruction here on earth.

After completion of *The Red Bomb*, I seriously considered producing a major multi-part series on the Soviet space programme, but it was a daunting task. Where to start? How to focus this enormous subject into a cohesive analysis?

A close friend in Moscow, one of my contacts within the KGB, solved the problem for me. One day he saw a Russian television programme making the silly and utterly unsubstantiated claim that Gagarin was still alive, having escaped from a mental institution where he had been incarcerated for the last thirty years. Even the KGB would not have been capable of 'disappearing' Russia's greatest hero. My contact urged me to make a serious documentary film that told the truth. At that moment it occurred to me that Gagarin's name was world-famous, yet nobody knew the first thing about him, or about the remarkable people who launched him into space aboard a converted nuclear missile. My next project was assured, in my own mind at least; and, as it happened, the BBC and other major broadcasters around the world shared my enthusiasm to reveal the unknown Yuri Gagarin.

Researching *The Red Bomb* took me and my production team into some very murky areas. For some reason I imagined that finding out about Gagarin would be slightly easier than unveiling the secrets of the ultra-secure Soviet nuclear-weapons programme. I was wrong. This time it was not so much the paranoid remnants of a massive State security system that I came up against, but the natural reticence of the individual Russian mind. The habit of discretion, at every level of society, has become utterly ingrained, long after the secret policemen have lost their power to enforce it.

I discovered that – even after all these years – many people who worked alongside Gagarin at the height of the Soviet space effort in the 1960s were unwilling to discuss their experiences, or to share their memories of the man himself. They were still afraid, I think, that some anonymous squad of men in heavy overcoats might turn up at their apartments in the early hours of the morning to arrest them for speaking out of turn. Perhaps more importantly, Westerners are still too foreign to be trusted with personal intimacies about old friends – and people who spent their entire careers in the Soviet military-industrial complex do not easily surrender their most sensitive technical secrets to outsiders.

Patiently, and I hope with tact and diplomacy, I persevered. I found that several key people were willing to speak, so long as they could give their testimony directly to fellow Russians rather than Western journalists. In this regard, my good friends Igor Morozov, Valerie Gorodetskaya and Maria Semenov were invaluable. They knew what was required and surpassed my greatest expectations. Morozov is a veteran of difficult, in-depth reporting from the old Soviet days. Gorodetskaya and Semenov have an astonishing ability to relax their interviewees while getting right down to the truth of the matter in hand. Gradually a degree of trust was established, and there was a 'snowball' effect, as reluctant witnesses took heart from the openness and enthusiasm of those who had spoken before them. An incredible amount of information emerged, and in this area we extend our thanks to Sergei Kuzhenko and Boris Malakhov for their invaluable help.

Inevitably the interviews had to be shortened, if the emerging television documentary was to fit into its standard 52-minute slot. I felt uneasy about wasting any of the stories entrusted to us by so many people who had been persuaded that we were going to make good use of their testimony. I decided that a book was needed, in addition to the film. Writer Piers Bizony joined our research team to co-write the book with me. It was a gamble, taken at short notice. Would we get along? Could we two very different characters, very much in control of our own different styles of storytelling and ways of working, collaborate on the same narrative without conflict?

Yes, we could. From the first, we saw exactly the same story, and the strength of our working relationship added to the strength of the book. I know something of Russia and its people, while Piers knows about their early rockets and the mind-set of the engineers who built them and the young pilots who flew aboard them. Together we explored fascinating and hitherto unknown aspects of Gagarin's life and work that surprised and moved both of us, and gave us an overwhelming sense that we were bringing to light a truly important story.

But no amount of teamwork could easily overcome the problems of searching for facts, documents and people in a country so vast and complex as Russia. The old bureaucracies no longer fly the hammer and sickle above their rooftops, yet they remain as impenetrable as ever. Money is now the dominant preoccupation. Certain things are possible for the right price. No one in particular is to blame for this; it is simply a basic fact of life in the 'New Russia'. The other problem is accountability. Who is in charge of any particular organization? Whom should we approach for access to buildings, for permission to use photographs or rights for film clips?

There are substantial historical archives in Russia, yet few people know exactly what they contain, because there simply is not the money available to catalogue and store them properly. Certain enterprising academics will search through them, for a price, but merely locating a particular resource does not guarantee access to it.

Take as an example what should have been a relatively simple task: obtaining the rights to reproduce a few minutes of archive footage of cosmonauts training for their flights. The space-medical institute in question demanded a horrendously high price for their footage, until their chief administrator agreed to release it for free, because he believed that his institute would benefit by having its work highlighted in our documentary. Then his deputy arrived back from vacation and overturned his decision. At last a senior advisor in President Yeltsin's administration cleared the way for us . . . until the space-medical institute complained, in no uncertain terms, that it was not answerable to Yeltsin – or anyone else in the government – and so the entire deal was off. We began our negotiations again, from scratch.

There was hardly a single instance where we contacted an interviewee without using intermediaries, or obtained documentary material directly from its source. Business relations in Russia are multi-layered and complex. Certain procedures have to be observed. Discovering the secrets of Gagarin's life and death came down to much more than simply picking up the phone and calling his old colleagues for a chat.

From our Western perspective, Russia is unimaginably vast. How could we possibly find the long-retired engineers who built Gagarin's spaceship, especially bearing in mind that their real names were kept secret throughout their careers? How could we identify and locate the humble farmworkers in deep rural locations (with no telephones in their homes), who saw Gagarin come home from space and land in a field? How could we even learn the names and addresses of the KGB escorts, long-retired and never identified at the time, who advised Gagarin during his global publicity tours in the mid-1960s? How to persuade the cosmonauts who vied with Gagarin for position in the space hierarchy to talk openly about things that simply are not supposed to be mentioned?

Once again, our Moscow researchers were indispensable to us in locating many of our witnesses and persuading them to speak. Some advisors within the modern, but still daunting, remnants of the KGB also helped us. They do not wish to reveal their names, but they know how grateful we are for their help in locating certain people.

But we owe our greatest debt to the interviewees themselves. As I hope this book will demonstrate, we have managed to speak to most of the surviving key figures in Gagarin's life. With considerable candour, occasionally at some risk, and almost always with a special Russian humour and emotion that brings their stories to life, the following people have entrusted us with their memories.

Gagarin's brother Valentin and sister Zoya told us about Yuri's early childhood and upbringing, as well as the impact of his famous space flight on the lives of those closest to him. They also narrated their family's terrifying experiences during the Nazi invasion of their homeland. These events shaped their brother's character. Yelena Alexandrovna, a retired schoolteacher, recalled Yuri as a bright, mischievous pupil, a merciless prankster, who also showed a deeper and more responsible side to his nature, remarkable for one so young.

Journalist Yaroslav Golovanov, who at one time actually trained for space flight himself, knew Gagarin and the other cosmonauts well. With his help we heard from Georgi Shonin and other Air Force pilots who were recruited into the space programme alongside Gagarin. Golovanov filled in the basic details of Gagarin's career, drawing on his encyclopaedic memory and his many books on the subject. Nevertheless, we know we have uncovered some important events that Golovanov himself may find surprising.

Sergei Belotserkovsky, a crucial figure in the academic training of all the early cosmonauts, knew Gagarin very well and provided us with valuable personal anecdotes and fascinating documentary evidence surrounding the investigation into Gagarin's tragically early death.

The diaries of the late Nikolai Kamanin, Chief of Cosmonaut Training, provided a close and unique insight into the character of the First Cosmonaut. Almost on a daily basis, Kamanin outlined the strengths and weaknesses of Gagarin and his great rival in the race to become the world's first spaceman, Gherman Titov. Through Kamanin's highly spontaneous scribbles, we saw just how close Gagarin came to being second. We are grateful to Kamanin's executors for permission to reproduce some extracts from a very colourful and outspoken journal.

For his part, Titov spoke frankly of the immense pain brought about by his role as Gagarin's back-up, or 'understudy', for the historic first space flight, while Alexei Leonov put many other aspects of a cosmonaut's life into perspective. These two men were very close to Gagarin, and they shared his experiences of flying into space and seeing the earth from orbit. They both had dramatic tales to tell.

Oleg Ivanovsky, Vladimir Yazdovsky and Yuri Mazzhorin – three of the most significant technical administrators in the Soviet space programme – revealed extraordinary details about their work, and of the spacecraft that Gagarin flew. They also revealed aspects of the deep humanity within the character of Sergei Korolev, the legendary 'Chief Designer' of Soviet rockets and spacecraft.

Sergei Nefyodov and Yevgeny Kiryushin, forgotten contributors to the space effort, recounted their secret work as 'testers', enduring great physical discomfort and risking death as part of the medical and physiological research programme that surrounded the early manned rocket programme.

Farm worker Yakov Lysenko, a very old man now, still gets a gleam in his eye as he recalls Gagarin landing in a field after his flight and greeting him; while Tamara Kuchalayeva and Tatiana Makaricheva recalled how, as schoolgirls, they ran across a gentle meadow to see where the world's first spaceship came to rest after its epic journey.

Anna Rumanseyeva recalled nursing Gagarin after a minor but embarrassing accident that nearly cost him his career. This was the first time that she, or anyone else, had told the truth about this significant and very human incident.

Sergei Yegupov, an archivist at the space training complex just outside Moscow, gave us access to some remarkable letters addressed to Gagarin from ordinary citizens, and illuminated for us some of the more difficult political aspects of Gagarin's career.

State security expert Nikolai Rubkin helped us analyse important details of Gagarin's fatal accident in a MiG training jet, and the flawed, compromised investigation that followed. Vyacheslav Bykovsky, an air-traffic controller, spoke to us about the day of the crash, even though this particular subject must have been very difficult for him. He had kept his silence for thirty years.

KGB veteran Venyamin Russayev came forward, after lengthy and delicate negotiations, to tell us an incredible and shattering story about Gagarin's efforts to save the life of Vladimir Komarov, the first man ever to die during a space flight. Russayev's evidence, revealed in our book for the very first time, is truly shocking and moving, and lays a trail all the way to the Kremlin.

Gagarin's wife, Valentina Gagarina, does not speak to journalists, but she was responsible for persuading Russayev to come forward to tell an important story that even now it is dangerous to reveal. We are deeply grateful for her blessing. Certain aspects of her husband's story must be difficult for her to dwell on. Conscious of this, we hope we have honoured Yuri Gagarin's memory and character, as well as telling the truth about his eventful and complex life.

We are indebted to Gagarin's personal driver Fyodor Dyemchuk for his memories, while Igor Khoklov, his favourite hairdresser, told us many fascinating tales. Khrushchev's speechwriter and senior aide Fyodor Burlatsky put Gagarin's intimate and difficult relationship with the Kremlin into perspective from close personal knowledge.

Of course this book is as much about the early Soviet manned space programme as it is about Gagarin himself, for the two stories – personal and technological – are intimately connected. Several Western experts in space history advised us about the engineering and administrative aspects of our narrative. Phillip Clark, Rex Hall, Brian Harvey and Gordon Hooper were endlessly patient with our questions. Douglas Millard at the National Museum of Science and Technology (London) provided many books and documents, while James Oberg and James Harford gave us advice on key issues and David Baker provided some excellent photographs. Andy Aldrin, John Logsdon and Peter Almquist outlined a particularly important aspect – the shocked American reaction to the first Soviet space triumphs. It is sobering to be reminded just how many exceptionally brilliant scientists and engineers were trained and employed within the Soviet regime.

Here, then, is the story of a moment that can never be repeated: the moment when one of our kind first ventured off the planet and into the cosmos. Many have followed, but only one man was first. Yuri Gagarin was no superman; he was mortal and flawed, just like the rest of us; yet he deserves his status in history: not just for the mere fact of being first into space, but also because he lived his life with decency, bravery and honour.

JAMIE DORAN

1

FARMBOY

This is the story of a young man who became famous in 1961, even though the world knew almost nothing about him. The great achievement of his life, celebrated to this day, took him less than two hours to complete, yet required bravery and commitment over a period of years. A triumphant superstar at the age of twenty-seven, he was tired, frightened and haunted by the time of his thirty-third birthday. In that last year of his short life he battled with his country's government to try and save a colleague destined for almost certain death; he met State security agents in darkened stairwells, avoiding hidden microphones, and passed on documents so sensitive that people could lose their jobs just for glancing at them. This man put his own life at risk, first for his country, then for his friends. Even his childhood required bravery, in the face of terrifying events that few of us could hope to survive.

We remember Yuri Alexeyevich Gagarin as the first person in history to travel into space, but there is much more to his life than this.

He was born on March 9, 1934 in the village of Klushino in the Smolensk region, 160 kilometres to the west of Moscow. His father, Alexei Ivanovich, and his mother, Anna Timofeyevna, worked on the local collective farm, he as a storesman, she with the dairy herd. Yuri's brother Valentin was ten years his senior, and a younger brother, Boris, was born in 1936. Despite hardships, the family was reasonably content, given the inevitable harsh conditions of Stalin's early collectivization programme and occasional unexplained disappearances among their friends and neighbours.

Responsibility for minding Boris and Yuri while Anna worked on the collective farm fell to the couple's only daughter, Zoya. 'I was seven when Yura was born, but at seven you already know how to be a nanny, so I got used to that. Of course, as a girl I was more responsible for looking after the littlest children, while Valentin helped out with the cattle on the farm.'

Official Soviet accounts of the Gagarin family as 'peasants' do not take into account Anna's origins in St Petersburg, where her father had worked as an oil-drilling technician, until the 1917 Revolution persuaded him to move his family into the country; nor the fact that she was highly literate, and never went to bed at night without first reading aloud to her children, or helping them to read for themselves.[1] As for Alexei, by all accounts he was a loyal husband, a strict but much-loved father and a skilled carpenter and craftsman, although there was a period in the early 1930s when it seemed best not to advertise his talents. Joseph Stalin had a murderous obsession with *kulak* farmers: anyone who made a reasonable living in agriculture or as a rural tradesman. When the collectivization programme became more firmly established, Alexei was made responsible for the maintenance of farm buildings and facilities, crude though they were.

At his side, the boy Yuri learned to tell the difference between pine and oak, maple and birch, just by the touch and smell of the wood. Even in the dark he could tell. His first experiences of materials, machinery and the technical possibilities of the world around him were bound up with wood shavings and the smooth feel of a good piece of carving; his early taste for precision, with his father's chisels, planes and saws.

Everything changed in the summer of 1941, when German divisions attacked the Soviet Union along a 3,000-kilometre front, making rapid advances against the Red Army. After several weeks of stunned inertia, Stalin's response was to order his divisions to pull back at each encounter, drawing the Germans so deep into Soviet territory that (like Napoleon before them) they were caught off-guard by the first Russian winter. The brief summer of Nazi success was followed, in essence, by a two-year retreat, with appalling casualties on both sides. The Smolensk region lay directly in the Nazis' retreating path. Gzhatsk and all its outlying villages, including Klushino, were overrun and occupied.

At the end of October 1942, German artillery units began to fire on Klushino. 'The front line was only six kilometres away, and shells were falling into our village every day,' Valentin recalls. 'The Germans must have thought the mill was a dangerous landmark, so they blew it up, along with the church. An hour later our own side launched an artillery attack in reply. It was all so pointless, because everybody must have had the same landmarks drawn on their maps.'

Soon after this barrage, four armoured German columns passed right through the village. There was a terrible battle in the surrounding woodlands, resulting in heavy casualties for both sides, but the Russian troops came off worst, with at least 250 dead or wounded. Two days after the fighting had subsided, the older Gagarin boys, Valentin and Yuri, sneaked into the woods to see what had happened. 'We saw a Russian colonel, badly wounded but still breathing after lying where he fell for two days and nights,' Valentin explains. 'The German officers went to where he was lying, in a bush, and he pretended to be blind. Some high-ranking officers tried to ask him questions, and he replied that he couldn't hear them very well, and asked them to lean down closer. So they came closer and bent right over him, and then he blew a grenade he'd hidden behind his back. No one survived.'

Valentin remembers Yuri's rapid transformation after this from a grinning little imp to a serious-minded boy, going down into the cellar to find bread, potatoes, milk and vegetables, and distributing them to refugees from other districts who were trudging through the village to escape the Germans. 'He smiled less frequently in those years, even though he was by nature a very happy child. I remember he seldom cried out at pain, or about all the terrible things around us. I think he only cried if his self-respect was hurt . . . Many of the traits of character that suited him in later years as a pilot and cosmonaut all developed around that time, during the war.'

Now the familiar tragedy of occupation came to Klushino: men in drab uniforms bashing down doors, dragging people away to be shot. If the need arose to preserve ammunition, they gouged at people with their bayonets or herded them into sheds and burned them alive, until the aggressors were broken in turn by their own misery, and ultimately by the cruel Russian winter and the unforgiving vastness of the landscape.

One particularly nasty piece of work, a red-haired Bavarian called 'Albert', collected the German vehicles' flat batteries in order to replenish them with acid and purified water, and also fixed radios or other pieces of equipment for the big Panzer battle tanks. Albert took an immediate dislike to the Gagarin boys because of their use of broken glass. The village children did what they could, smashing bottles and scattering the bright shards of glass along the roads and dirt tracks, then hiding in the hedges to watch the German supply trucks swerving out of control as their tyres burst. Albert became convinced that Boris was one of these child-saboteurs. He came across the boy playing with Yuri, and sat down on a nearby bench to watch them. After a while he offered Boris some chocolate, putting it on the ground so that when the boy reached for it, he could stamp on his fingers. 'The skin came right off his fingers, so of course Boris cried out,' says Valentin. 'Then the Devil took him – we always called him the Devil – and hanged him by his scarf on the branch of an apple tree. Mother came and found the Devil taking pictures with his camera. It's difficult to talk about . . .' Anna scuffled with the German, and at one point he picked up his rifle. For a terrible moment it seemed as if he was ready to shoot, but by some miracle one of his superiors shouted to him to come away. Fortunately, the Devil's work had been sloppy, and a child's woollen scarf did not make a very effective noose. Once Albert was safely out of the way, Anna and Alexei released Boris from the apple tree.

Albert and his fellow soldiers had thrown the Gagarins out of their home, and the family had been forced to dig for themselves a crude shelter in the ground. Here they carried Boris's limp body, and by sheer force of will and desperation they brought their throttled child back to life. 'Boris stayed in the dug-out for a week, terrified to go out,' says Valentin. He also remembers that Albert came across a rare Gagarin family luxury in the house, a wind-up gramophone, and played a particular record time and again, hoping to taunt the Gagarin family as they huddled in their rough shelter. 'He would open the window of our house and play the military march "Red Army Advance" as loud as he could. Obviously he didn't know what it was.'

In the days after the terrifying drama by the apple tree, Yuri began a ceaseless vigil, watching for the times when Albert would leave the house. Whenever it was safe, he crept over to the Germans' precious pile of tank batteries and dropped handfuls of dry soil into the accumulator caps to ruin them, or muddled up the chemical replenishment stocks, pouring them willy-nilly into the wrong compartments. Albert and his companions would get back to find their batteries looking perfectly normal, and patrolling tank drivers would arrive in the mornings to pick them up. They would shake Albert's hand, make their Nazi salutes and be on their way, but at evenfall they would return, furious that Albert had given them dead batteries. Most of the tank commanders were SS officers, so their displeasure was a very serious business for everyone, German and Russian alike. 'They were very hard to pacify,' Valentin recalls drily.

Humiliated by the anger of the SS officers, Albert went on the rampage, searching the village for Yuri, but he had to hunt on foot because his child-nemesis had shoved potatoes deep into the exhaust pipe of his military car, so that it would not start. The Devil stormed his way into all the dug-outs, threatening to shoot Yuri on sight. Perhaps the German commanders were impatient with Albert's dead batteries by now, because they called him away from the district before he could finish the boy off once and for all.

Valentin was placed in a work detail by the Germans, along with eight other lads. 'The rules were very simple. You started work at eight in the morning, and then you could die, or else you worked until they said to finish. Even if you were halfway through chopping a tree and it was about to fall on your head, you had to stop the instant they told you, or else you'd feel a stick or a rifle butt.' As the Germans began to dig in for the winter, pretty much trying to survive, like the villagers, occasional confusions developed as to who was the enemy, who the aggressor. There was one particularly large communal dug-out, capable of supporting three or four hundred people, but whether this was a German or a Russian construction nobody could say, since it was built and used by both sides at once. Valentin recalls, 'Somebody's aircraft attacked it one morning, dropping a clutch of bombs onto it – a tonne and a half each, the Germans reckoned. No one knew for sure how many were killed.'

During the spring of 1943, Valentin and Zoya were abducted by SS guards and herded onto a 'children's train' for deportation to Germany. They were taken first to Gdansk, in Poland, where they worked in adjoining labour camps. 'I had to do the washing for hundreds of Germans each week,' Zoya says. 'We lived as best we could, but they were the proprietors and we were the slaves. They could have done anything they liked to us – killed us, or let us live. We were worn down with fear all the time, and we looked like ragged Cinderellas, all skin and bone, with our elbows sticking out. We had no shoes, and occasionally found soldier's boots that were too big for us . . . The Germans put us in ruined houses after they'd expelled the people already living in them.' Zoya does not like to dwell on her experiences as a 15-year-old girl hauled away by enemies.

In the chaos of the Germans' long retreat from Russia, the SS use of trains for prisoners was considered something of a luxury by the ordinary troops. The 'children's trains' running through Poland were commandeered or otherwise diverted from their original course. Valentin and Zoya escaped their camps and spent two weeks hiding in the woods, waiting for Russian troops to rescue them. 'When they actually came, we hoped they would let us go home,' Zoya recalls, 'but they said we must stay with the Russian army as volunteers.' Zoya was sent to look after horses in a cavalry brigade and, by a bitter stroke of irony, followed them deep into Germany, where the children's train was supposed to have taken her in the first place. By now,

Valentin was considered old enough for front-line service. He quickly learned how to handle an anti-tank grenade launcher and other heavy weapons.

Meanwhile, Alexei and Anna Gagarin thought their two eldest children were dead. Alexei, never a very fit man, was ulcerated with grief and hunger, and was seriously injured when the Germans beat him up when he refused to work for them. He spent the rest of the war in a crude hospital, first as a patient, then as an orderly. Anna spent some time there too, with her left leg badly gashed after a German sergeant, 'Bruno', had flailed at her with a scythe. Yuri threw clods of earth in Bruno's eyes to drive him away.

The Germans were driven out of Klushino at last on March 9, 1944. Alexei, limping but defiant, showed the incoming Russian forces where the fleeing Nazis had buried mines in the surrounding roads and dirt tracks. Anna recovered from her wound, and struggled to look after Boris and Yuri, although there was almost no food of any kind to be had. Only towards the end of 1945 did she discover that Valentin and Zoya were still alive. They came home at last, grown-up now.

Lydia Obukhova, a writer who came to know the Gagarins well during the 1960s, commented in 1978:

> Valentin was still a boy, and Zoya was a young and charming lass, defenceless in the face of misfortunes that might befall her far from home. Her mother's grief was boundless, but her husband said to her, 'Remember, Boris and Yuri still need you.' You'd have thought the war, the occupation, the fearful Germans billeted in the Gagarins' home, would have mutilated for ever those children's personalities, but their mother and father did everything to prevent this. They never showed even a trace of servility to the enemy. It follows that the children showed none either.[2]

After the war, the Gagarins moved to nearby Gzhatsk and built a simple new home, using the slats and beams from the wreckage of their old house as raw material. The original house had been very modest anyway, consisting only of a kitchen and two small adjoining rooms. 'Of course life was hard after the war,' Zoya explains. 'Everything from Brest up to Moscow was completely destroyed, all the cattle taken away, the houses in ruins. There were only two houses left in our village.' The people of Gzhatsk built a school, and a young woman, Yelena Alexandrovna, volunteered to run it, making the best of a blackboard with no chalk and a classroom with no books. Yuri and Boris learned to read from an old Russian military manual left behind by departing troops. For geography they relied on war maps rescued from the burned-out cabins of army trucks and tanks.

Yelena was not on her own at the school for long. In 1946, Lev Mikhailovich Bespavlov joined the school to teach maths and physics. A new father figure had now arrived in Yuri's life. Speaking to an Australian journalist in 1961, Yuri described Bespavlov as 'a wizard, specially when he'd fill up a bottle with water, and seal it, then take it out into the freezing air outside, so that the water would turn to ice and expand, shattering the bottle with a satisfying bang. Bespavlov could float pins on water, and make electricity by combing his hair.'[3] Perhaps the greater part of his appeal lay in the faded airman's tunic he sported, for in the chaos and terror of the war years, Yuri had encountered one thing so wonderful, so magical, that it seemed for a moment to transcend the horror all around – an aircraft; and even when this piece of magic had been dismembered and taken away, its memory remained.

There had been a dogfight, two Soviet 'Yak' fighters, two German Messerschmitts, with the score levelling out at one-all. The stricken Yak came down in a patch of marshland half a kilometre outside the village. One of its landing legs buckled on impact, and the propeller was twisted completely out of shape. The ground was soft, which made for a very poor landing, and although the pilot survived, he grazed his leg quite badly. Immediately, a crowd of villagers ran across to help him. They put a bandage on his injured leg, offered him a drink of milk and fed him some pieces of dried bacon.

After a while another Russian aircraft, a Polikarpov PO-2, came down safely in an adjacent clover field with firmer ground. Airmen called the PO-2 a 'cornplanter', because its lightweight plywood construction enabled it to make landings in rough fields. Today its apparent 'rescue' mission was somewhat double-edged; the PO-2 crewman was supposed to check on the health of the downed Yak pilot, then ensure that his fighter did not fall into German hands, if necessary by destroying it.

Yuri watched all of this, mesmerized. According to Valentin, 'Some of the older boys in the village were sent into the clover field with whatever dregs of petrol they could scavenge, to refuel the PO-2. The pilot had some bars of chocolate, which he gave to Yuri. He divided them among several other boys, accidentally keeping none for himself, obviously much more interested in the planes.'

As the light faded, the two pilots were invited to shelter in a dug-out, but chose instead to spend the night huddled near the PO-2 to keep watch over it. They tried to keep guard throughout the night. Inevitably, cold and bruised, they fell asleep and awoke early next morning to find Yuri staring at them. In the light of day, the damaged Yak fighter did not really seem worth guarding any more, so the pilots set fire to it, then struggled back over the fields to the PO-2, the injured pilot leaning on the other's shoulder for support. They coaxed the 'cornplanter' into the sky without too much trouble and flew away, while Yuri watched, fascinated, as a tall column of smoke billowed from the wreck they had left behind.

Now the boy's teacher, Lev Bespavlov, carried with him some of that special magic in his uniform, which he had rightfully earned as a gunner and radio operator in the Red Army Air Force. Yuri looked up to him, listened and learned.

Yelena recalls Yuri being a good pupil: mischievous but honourable. 'Like all children of that age, he did some naughty things, but if ever we were asking the pupils, "Who did it?", Yuri would always say, "It was me, I won't do it again." And he was very vivid. Recalling those years, I would say he was a very decent and responsive boy. When we learned about his flight into space we immediately remembered his very nice smile. He preserved the

19

same smile for the rest of his life – the same one he had when he was a boy.' Yelena remembers placing Yuri at the front of the class for a few days, where she could keep an eye on him. 'He wasn't really the sort of boy you could take your eye off for too long. Even right under my nose, he managed to find trouble. He pulled all the nails out of the bench at the front, so that when he and the other children sat on it, the whole thing collapsed.' But Yelena could not stay annoyed for long. She remembers a tiny little girl, Anna, who kept getting trampled or left behind when the other children stampeded about the place. Yuri became quite protective of her; carried Anna's satchel after school and walked her home, to show the others that she should not be picked on.

Unfortunately, he did not shine at music. 'He participated in all the amateur talent activities. The instruments for the orchestra were a present from the collective farm. Yura played trumpet. He was always proudly walking in the front.' The Gagarin family had to survive, rather than enjoy, these atonal outbursts, as Zoya remembers. 'He brought his trumpet home and started to practise. Father got fed up. It was a sunny spring day, and Father sent him outside, saying he had a headache because of the noise. So he practised outside. We had a cow, and she started to moo. It was a concert for free. Everyone was laughing.' Zoya fondly recalls her younger brother as 'a real live wire. He was always leading games, the instigator rather than the follower. He was very much alive.'

Yuri's favourite subjects at school were maths and physics, and he was also keenly involved in a model aeroplane group, much to Yelena's inconvenience. 'Once they launched one of his planes from a window and it fell on a passer-by. He was exasperated, and came into the school to complain. Everyone went very quiet, until Yuri stood up and apologized. So he probably had this urge to fly.'

Valentin remembers his pesky brother at six years old, demanding that he and his father build him miniature gliders, or wooden propeller toys powered by rubber bands. Little Yurochka would insist, 'I want to be a hero for my country, flying a plane!' Until the war, at least, planes were seldom seen in the skies above Klushino. Fleeting glimpses of such craft must have made a powerful impression on the boy.

When Yuri was sixteen he became anxious to get away from home and earn some kind of living. 'He saw that life was very hard for our parents, and he wanted to get a profession as soon as possible, so that he wasn't a burden on their shoulders,' says Zoya. 'Personally, I didn't want him to go, but he said he wanted to carry on studying, and our mother said she wanted him to study, too.' Yuri expressed his enthusiasm for the College of Physical Culture in Leningrad. He was a fit young man, not very tall, but agile and coordinated. He thought he might train as a gymnast or sportsman. Valentin remembers their father's objection to this plan. 'He said it was not a job. Even though it might be physically hard work, it was a silly thing to do. But the physics teacher, Bespavlov, insisted that our parents let Yuri go.' Alexei hoped that one day his three sons would join him as carpenters, but such a plan was not really practical.

In the event, all the Leningrad places were taken. The best available option was at the Lyubertsy Steel Plant in Moscow, which incorporated a school for apprentices. Here Yuri could learn a proper trade – steel foundryman. There was much pulling of strings with relatives on his father's side of the family for interviews, references, accommodation. In 1950 Yuri was finally accepted as an apprentice and went off to Moscow, where Uncle Savely Ivanovich agreed to let him stay with them for a while.

At Lyubertsy, Yuri Alexeyevich Gagarin received his first adult uniform: a foundryman's peaked cap with a union emblem; a baggy tunic of cardboard-stiff serge, with the sleeves much too long; dark baggy trousers; and a wide leather belt with a big brass buckle. He looked at himself in the mirror and decided that his comical outfit was worth a picture. He spent his last few roubles sending a photograph back home.

Gagarin's foreman at Lyubertsy, Vladimir Gorinshtein, was a dour, heavy-set man with a drooping moustache and bulging muscles, and a tongue on him as scalding as the molten steel he so loved to work with. 'Get used to handling fire,' he would say to his cowering apprentices. 'Fire is strong, water is stronger than fire, the earth stronger than water, but man is the strongest of all!'[4] 'We were scared,' Gagarin recalled in a 1961 interview.

His first assignment was to insert hinge-pins into the lids of newly assembled metal flasks. The walrus-faced foreman strode across to inspect the work. By beating his fists against his forehead and swearing mightily, he was able to hint that Gagarin had installed his pins completely the wrong way round. 'The next day we all made better progress,' Gagarin recalled. By his own admission, this was typical for him. He had no particular knack for getting things right the first time. He had to work quite hard at his tasks, practising them repeatedly. In a brief interview given many years later, Gorinshtein said:

> At first Yura struck me as too small and frail. The only vacancy I had available was in the foundry group, which meant a lot of smoke, dust, heat and heavy lifting. I thought it would be beyond him. I can't remember why I eventually ignored all these negative points and accepted him. It must have been the determination you could feel in him. Was he special? No, but he was hard-working.[5]

Gagarin's year-end report from the foreman was good. In fact, he was one of only four apprentices to be selected for training at a newly built Technical School in Saratov, a city port on the great Volga river. Here he would learn the intimate secrets of Russia's most important machine: the tractor.

In the spring of 1951 Yuri and his three lucky companions from Lyubertsy were escorted to Saratov by their new teacher, Timofei Nikiforov. Within a few hours of their arrival in the town, Gagarin saw a notice. 'AeroClub' it read. 'Ah, my friends. That would be something. To get in there!' His companions laughed, but a few days later the club accepted his application to join. To his dismay, Gagarin found that the Technical School kept him relentlessly occupied, and it was several weeks before he could actually go to the club's airfield on the outskirts of Saratov.

Dmitry Martyanov, the club's war-veteran chief of training, saw Gagarin for the first time as a young man with a rapturous expression on his face gazing at an old canvas-clad Yak-18 training plane, so he strolled across and offered to take him for a brief trip into the air. They went up to 1,500 metres, crawled through the sky at 100 kmph, and came back down to earth after a few minutes. 'That first flight filled me with pride, and gave meaning to my whole life,' Gagarin recalled.

Martyanov said, 'You handled that very well. One would think you'd done this before.'

'Oh, I've been flying all my life,' Gagarin replied.[6]

Apparently Martyanov knew exactly what he meant, and he became a firm friend from that moment.

In the spring of 1955 21-year-old Yuri graduated with an 'excellent' grade from Saratov Technical School. By this time his interest in tractors was waning. He had spent the previous summer at the AeroClub learning to fly the Yak-18. After his first solo jaunt, he gave his friend and tutor Martyanov a pack of Troika cigarettes, a sort of traditional pilot's gift. But it wasn't all fun and games. He had to attend evening lectures in aviation theory, while keeping awake in the daytime for Technical School. He ploughed through the extra workload, determined not to fail. His reward at the club was to make a hair-raising parachute jump from the wing of a plane. Martyanov also recommended Gagarin for the Pilots' School at Orenburg, on the Ural River. Of course Gagarin would have to sign up as a military cadet, if he wanted to get in there. Orenburg was not some cosy little 'club', but a deadly-serious training centre for military fliers; though it has to be said that the AeroClub wasn't exactly for fun, either. In Soviet society, 'fun' was a difficult concept to grasp. Better to say: at the AeroClub eligible citizens could volunteer informally to practise a useful skill.

At the very least, 'fun' was supposed to keep you fit for work. Gagarin signed up with various sports clubs in Saratov and built up his small, undernourished frame with plenty of exercise and food. He played volleyball and basketball, and showed off to pretty girls sitting on the banks of the Volga how he could water-ski on one leg. If he fell off and clambered to shore wet and grinning, that seemed to impress them just the same. He had grown into a very fine-looking and confident young man, albeit slightly shorter than average and better suited for acrobatic rather than sprint sports. His good-natured charm and generous humour won him many friends.

The tutors at Orenburg were not so easily charmed. They were soldiers on active duty. Gagarin would have to commit to their discipline for years to come and, after they had finished with him, they might send him away to fight and be killed. Anna and Alexei Gagarin expressed dismay at the thought of their boy enlisting with the military, and this obsession with aircraft seemed reckless. However, Alexander Sidorov, a fellow member of the Saratov AeroClub, recalled in 1978 that Gagarin already had the keen self-discipline expected of a future military pilot:

> On overnight stays at the Saratov airfield, Yuri would zealously ensure that our camp tent was kept perfectly tidy. He couldn't bear untidy or slovenly people. At first he'd appeal to their consciences in a friendly way, but if that didn't work he would demand in stronger terms.[7]

It was not easy to score top marks at Orenburg. Yadkar Akbulatov, a senior instructor, said in 1961, 'Don't imagine that Yuri was an infallible cadet, a child prodigy. He wasn't. He was an impetuous, enthusiastic young man who made the same slips as any other.' His worst marks were for his landings. He was in danger of failing Orenburg completely if he could not get his aircraft down without bouncing on his tyres. Akbulatov flew with him a couple of times to see if they could iron out some faults. 'I took him up and watched him carefully. On steep banking turns his performance wasn't absolutely perfect, but in vertical dives and climbs he put on a show that made me see stars from the g-load. Then came the touchdown. It was faultless! I asked him, "Why can't you always land like that?" He grinned and said, "I've found the solution." He put a cushion under his seat so that he could get a better line of sight with the runway.' From now on, Gagarin never flew any aircraft without his cushion.[8]

True to form, Gagarin had worked on the landing problem with bloody-minded determination and had solved it, cushion and all. Now for the target practice. After all, what was the use of a military pilot who could not shoot the guns on his plane? At first, his practice gunnery strafes fell completely wide of the target, while his classmates at Orenburg scored bull's-eyes. Gagarin re-sat all the theoretical classes on the ground, tried again from the air and eventually destroyed his targets in the approved manner.

One notable sporting incident boosted Gagarin's career prospects while he was captaining the Junior Cadets' basketball team. After one particular game, where they thrashed the opposition, the tutors at Orenburg praised him for his skilful play. 'We didn't win because we played better,' he said. 'We won by sheer determination. We were bent on winning, while the other side hadn't made up their minds.' This statement impressed some of the senior officers watching the game. Akbulatov and his colleagues began to think of Gagarin as a contender. No genius, but a winner for all that. Plus, he was a young man who loved pulling heavy g's, an appropriate enthusiasm for a would-be fighter pilot.

At Orenburg, Gagarin met Valentina 'Valya' Goryacheva, a pretty, hazel-eyed medical technician one year younger than him. She worked on the Orenburg base, and came to a dance party one evening, only to find that the callow cadets with their short, bristly haircuts did not seem particularly impressive. In an interview with journalist Yaroslav Golovanov in 1978, she recalled that the civilian boys in downtown Orenburg seemed better dressed, had nicer hair and were more handsome. She never expected to find a love match at a military compound, just a pleasant night's partying. She danced with Gagarin a couple of times, while he cheerfully asked questions about her.[9]

At ten o'clock precisely the music stopped. The cadets were

expected to go to bed now (alone), so that they would be ready for an early start the next morning. Gagarin said, 'Well. See you next Sunday.' Valya made no reply. 'Back home, I thought: Why should I go and meet that bald-headed character again? In any case, why does he behave in such an assured way? But the next Sunday we went to the cinema. We had a difference of opinion about what we saw. Then afterwards he said, "Well. See you next Sunday."'

And he did. Valya's parents, Ivan Goryachev and Varvara Semyonova, lived in Chicherin Street, Orenburg. Valya had never known any other home town, but for Gagarin this place was entirely new and unknown. Valya's parents, along with her three brothers and three sisters, quickly became fond of him and their house became a kind of second home. Ivan cooked for the local sanatorium and applied his considerable chef's talent at home. No great fan of the dull food at the flying school, Gagarin ate well during his many off-duty visits to Chicherin Street. According to Valya, when she was interviewed in 1978, there was only one serious problem on Gagarin's mind at that time: 'His parents were having a hard time making ends meet, but how could Yura help them in any way from his small cadet's pay? He said it would have been better if he had gone back to Gzhatsk as soon as he'd finished at the Technical School so that he could earn a living in the profession he had learned, as a foundryman.' But Gagarin persisted with his training at Orenburg, achieving the rank of Sergeant in February 1956, and making his first solo flight in a MiG-15 jet on March 26, 1957.

On October 4, 1957, the Soviets launched the world's first artificial satellite. The cadets at Orenburg rushed around in great excitement when they heard the news. Gagarin's best friend on the base, his namesake Yuri Dergunov, ran towards him on the tarmac shouting 'Sputnik!' at the top of his voice. Gagarin was excited too, but the moment did not immediately change his life, as some accounts have suggested. He was much more concerned with his pending final exams at the Pilots' School, and with his growing love for Valya. The date for their marriage was already set for October 27. He recalled that at the time the wedding arrangements seemed far more pressing to him than any thoughts of flying into space; besides, the idea of sending people into the cosmos still seemed a distant abstraction. He never imagined that he would be in orbit in three-and-a-half years' time.

Valya had no idea that rocket travel would feature in her husband's life. She married a charming but essentially ordinary young military flier, not some future space hero. She must have known the risks involved, even in this simple relationship: that she would end up moving from one strange town to another as Yuri moved between different air stations; that he might set out to work one morning and not come back in the evening . . . Obviously she made these accommodations. She was entitled to expect that the military would assist with housing, health care, pensions and schooling for any children she and Yuri might raise. In return, she knew that she might have to grieve silently and without fuss if her husband was killed flying in his jet plane. Many other wives in her position shared the same burdens, the same nagging fears, and to some extent this must have helped. Valya made several close friends among the pilots' wives and, later, among the wives of the cosmonauts. What she never anticipated was becoming the wife of the World's Most Famous Man. She was a shy and intensely private young woman.

Nor was Valya thrilled with her new husband's first proper posting after he passed out of Orenburg with excellent grades and a Lieutenant's commission on November 6, 1957. Shortly after graduating, Gagarin was sent to the Nikel airbase on the northernmost tip of Murmansk, 300 kilometres north of the Arctic Circle, with an assignment to fly MiG-15 jets on reconnaissance missions. Valya followed him out there and discovered a terrible hinterland of sub-zero temperatures, biting winds and long, pitch-black nights, interspersed by a few hours of gloomy grey daylight. Here, on April 10, 1959, she gave birth to the couple's first daughter, Lena.

Throughout the long winter months flying conditions at Nikel were awful. Ice menaced the control surfaces of Gagarin's MiG, and snow blindness was a constant threat, with the sky and the ground merging into a seamless sheet of white with no discernible horizon. The on-board electronic approach and landing systems of a MiG were not particularly sensitive in those days. Snow-blinded pilots had to rely on the bigger ground-control radars to guide them towards the narrow radio beacons on the perimeter of the runway. Even in clear weather there were hazards. One day Gagarin put his plane down on a landing strip covered in black ice, transparent to the naked eye but slippery as oil. He had never practised at Orenburg for these conditions. His plane skidded violently and the landing gear's tyres burst under the strain of his sudden braking.

Gagarin's good friend Yuri Dergunov from the Orenburg Pilots' School had campaigned hard to be assigned the same posting when he qualified. It was a great shock when he crashed and was killed in his first month at Nikel. Valya recalled, 'For several weeks Yuri walked around in a daze and spent one sleepless night after another. I knew that no relaxing draughts or sleeping pills would help him, and if I offered him medicine it was only to take his mind off his depressing thoughts for a few moments at least.'[10]

2

RECRUITMENT

In October 1959 mysterious recruiting teams arrived without warning at all the major air stations in the Soviet Union, including Nikel. They did not say exactly what they wanted, or which organization they represented. Groups of pilots were selected and summoned into an office, twenty or so at a time, for an informal conversation with some 'doctors'. A few days later the requested groups became smaller, the many rejects winnowed out through mysterious consultations in the background, until eventually the recruiters were holding private interviews with just one candidate at a time, from a shortlist of perhaps a dozen from any given airbase. These lucky ones were sent to the Bordenko Military Hospital in Moscow for a series of rigorous health checks. In all, nine out of ten candidates failed this stage of the review process, again because of secret decisions taken behind closed doors.

In his published accounts, Gagarin recalled undergoing seven separate eye examinations in the Bordenko hospital, countless interviews with psychologists and a nightmarish mathematical test, during which a soft voice whispered all the solutions to him – the wrong solutions – through a pair of headphones. He had to concentrate on his own thought processes and ignore 'the obsequious friend' whispering so helpfully in his ears. The doctors were 'as stern as State prosecutors. Our hearts were the main object of their examination. They could read our whole life history from them. You couldn't hide a single thing. They tapped our bodies with hammers, twisted us about on special devices and checked the vestibular organs in our ears. We were tested from head to toe. Complicated

29

instruments detected everything, even the tiniest cracks in our health.'[1]

The exact dialogue between Gagarin and his recruiters is not on record. However, one of his fellow pilots at the Nikel airbase, Georgi Shonin, was assigned for cosmonaut training soon after Gagarin, and his experience of the selection process must have been pretty similar. 'At first we talked about the usual boring things. How did I like the Air Force? Did I like flying? What did I do in my free time? What did I like to read? And so on. A couple of days later a second round of talks began. This time far fewer of us had been called for. The talks now became more specific.' Shonin went to Moscow without really knowing why, then came back to Nikel utterly punctured, probed, exhausted, but apparently listed fit by the doctors for whatever new job awaited him. The recruiters sat him down one last time. 'They asked me, "How would you feel about flying in more modern planes?" And the answer dawned on me.' At that time some air regiments were switching over to new high-performance fighters and Shonin assumed that the Nikel squadron was about to become one of the lucky ones. But no, the recruiters were not interested in anything so banal as putting him in a new kind of MiG. 'They asked me, "What if it was a question of flying something completely new?" I immediately cooled off, because I knew a lot of pilots were being transferred to helicopter units.' Shonin regarded helicopters as complex but unappealing machines with no speed, no style and, above all, no status. 'I'm a fighter pilot,' he pleaded. 'I specially chose a flying school where I'd train for combat jets, and now you want me –'

One of the recruiters interrupted him. 'No, you don't understand. What we're talking about are long-distance flights on rockets. Flights all the way round the earth.'

'My mouth dropped open in surprise,' Shonin remembers. But he took the job. Gagarin was recruited ahead of him by several weeks, and although no record exists to prove it, he and Valya must have welcomed a chance to escape from Murmansk with honour intact. He explained to her that he had been selected as a test pilot for new types of aircraft and was to be stationed just outside Moscow. They left Nikel on March 8, 1960, gladly giving away their standard-issue furniture to other families on the base.

On starting his new job, 26-year-old Gagarin found himself in a group of just twenty 'cosmonauts' finally selected out of an initial candidate list of 2,200 from all over the Soviet Union. This first squad would become very tight-knit, despite the obvious (if unspoken) rivalry to gain actual space flights. Gagarin's relationship with fellow recruit Gherman Stepanovich Titov would develop into something much more complicated – an unspoken contest to win the first human flight into space.

To the other cosmonauts, Gagarin came across as an easy-going fellow. By contrast, Gherman Titov seemed proud and aloof, even to his friends, and sometimes rather strange, too. He loved to spout reams of poetry or quote fragments of stories and novels, not just the modern approved works, but real literature from the old Tsarist days. His father, Stepan Pavlovich, was a teacher and had named him Gherman after a character in a story by Alexander Pushkin, 'The Queen of Spades'.[2]

Titov's was an intensely self-possessed character. In 1950, at the age of fourteen, he crashed his bicycle and broke his wrist, but instead of running home and pleading for comfort, he pretended that nothing had happened. He nursed his pain in secret, unwilling to admit any weakness because he had already signed up for elementary training in aviation school at the next available opportunity, and he did not want his accident to spoil his chances of selection. When he finally received his papers as a cadet in 1953, he worried that the military medics might investigate his bones and call his bluff. They did not; instead, he called theirs. He took to performing early-morning exercises on a set of parallel bars, until his flawed wrist appeared as good as the other. He trained at the Volgograd Air Station for two years, graduating with distinction, and in 1959 was, like Gagarin, interrogated by mysterious visitors as a potential cosmonaut. The space doctors took Titov apart with greater rigour than the Air Force, but still they failed to find anything amiss in their X-rays, and he was selected to be part of the first group of twenty cosmonauts, along with Gagarin. Long after it was too late, the medical examiners told him that if they had known about his old wrist injury, they would never have sanctioned his recruitment.

While Gagarin's face was open and quite easy to read, Titov's eyes were hooded and dark, his nose set stern. His mouth often seemed to narrow in disapproval, so that his expression could verge on arrogance. He was quick-tempered and was not afraid to speak his mind. He wore his uniform smartly, and his glossy, brown wavy hair helped create the impression of a bourgeois cavalry officer; as did his personal pride, too much poetry and a suspicion of class (just because he had read a few books and his father was a schoolteacher). Titov was not the sort of man who would easily prosper in an egalitarian workers' and peasants' paradise – except for one valuable saving grace. He had proved himself excellent in one of the few realms of Soviet life where individual excellence was encouraged: up in the air, protecting the Motherland in a MiG.

Alexei Arkhipovich Leonov was another important figure in Gagarin's new life as a cosmonaut. Born in May 1934, Leonov was almost exactly the same age, although his well-rounded, happy face and slightly thinning hair gave him a more middle-aged appearance. As a young man he considered becoming an artist, and in 1953 he enrolled at the Academy of Arts in Riga, but almost at once he had a change of heart, applying instead to the Air Force school at Chuguyev. There he became a talented parachutist and qualified as an instructor. His chunky features belied a super-fit condition, and he trained with intense self-discipline as a fencer and runner, almost always finding time in the early morning for a four-kilometre workout. When the space recruiters came to Chuguyev in 1959 he soon got himself a new posting.[3]

Leonov's sense of humour was infectious, often mischievous, and as long as he was not thwarted or irritated, he was a genuinely lovable man, although in keeping with all his cosmonaut colleagues he was fiercely determined to succeed in his new profession. When it came to work and training he could be cold-blooded at times, but by and large his humour and professionalism won him many allies. He always retained his fascination for art, taking sketchbooks everywhere he went (including into orbit) and eventually becoming one of the leading Soviet space artists. He became Gagarin's closest friend from the earliest days of their cosmonaut training. 'I quickly discovered for myself the kindness of this man's nature,' Leonov says today. 'Yuri liked his friends very much and paid them a lot of attention. He kept in touch with old ones and very easily made new ones. Our relationship was especially warm because we knew each other for a long time and, even when he became famous, it didn't affect him. He always stayed a very good friend.'

On January 11, 1960, a special cosmonauts' training centre was inaugurated under the directorship of medical scientist Yevgeny Karpov. His second-in-command, directly responsible for the cosmonauts' recruitment, training and 'ideological reliability', was General Nikolai Kamanin, a tough and highly ambitious combat veteran with no discernible sense of humour. The space historian James Oberg has described him as 'an ageing war hero and authoritarian space tsar, a martinet',[4] while Yaroslav Golovanov, who was intimately connected with the early Soviet space programme, remembers him as 'a terrifyingly evil man, a malevolent person, a complete Stalinist bastard'. In time, many of the cosmonauts would come to hate Kamanin, but his strict military discipline, his remorseless attention to detail and his refusal to accept anything but the highest standards from his students would prepare them successfully (in most cases) for the rigours of space.

When Gagarin and his nineteen colleagues arrived for training, few suitable facilities existed to prepare them for space. Karpov and Kamanin were allocated a large swathe of birch and pine woodland forty kilometres north-east of Moscow, and in March 1960 they began construction of a secret base called *Zvyozdny Gorodok*, or 'Star City'. A huge square compound was cleared in the middle of the site, thoroughly screened from any nearby roads by the surrounding woodlands. A simple hostel was built; some standard-issue barracks; and a number of low buildings to house the training facilities, some of which – as the cosmonauts would discover to their cost – were designed to inflict stress, trauma, loneliness and exhaustion. In time, Star City would grow to the size of a small town, with its own private community of bars, hotels, sports clubs and administration centres. A short distance to the south, a large and sprawling Air Force base at Chkalovksy provided convenient landing strips for jet trainers and cargo planes, as well as accommodation for the cosmonauts and their young families.

Despite the size of the Star City construction zone, few people outside the space effort knew very much about it. The road that passes Chkalovsky skirts the dense pine forest that so effectively conceals the complex. On the right-hand side a small guardpost protects an innocent gap in the mask of trees. In 1960 the turn-off road might easily have been mistaken for a loggers' track or an old farm route, except for the guardhouse and the solid tarmac surface capable of taking heavy trucks.

When Gagarin and his colleagues were recruited, the facilities at Star City were not yet operational. The cosmonauts' early training consisted mainly of academic and physical work in various Moscow scientific and medical institutions, in particular at the Zhukovsky Academy of Aeronautical Sciences on Leningradsky Prospekt. The least popular venue was the Institute for Medical and Biological Problems near Petrovsky Park, headed by Oleg Gazenko, where all the cosmonauts underwent a bewildering array of medical, physical and psychological tests. One of the procedures they all had to face here was the 'isolation chamber', a large sealed tank capped by an airlock and containing the barest of living accommodations. The doctors would seal their victims inside, then raise or lower the interior air pressure according to their scientific whims. Then they would provide a miserable set of tasks: maths sums, intelligence tests, physical exercises, and so forth. What the inhabitants of the chamber were not allowed to do was pass the time by somehow enjoying themselves. No chatter was permitted; no books, no magazines; no contact with the outside world, except for the most minimal dialogue with the technicians monitoring the chamber. A session might last anything from one to ten days, although the victim was never warned in advance how long it might be. The purpose of this ordeal was to determine if a man could survive the boredom and loneliness of life in a spaceship, perhaps held in orbit above the earth for many days during some unfortunate delay in the re-entry sequence. That was the official explanation for the chamber. For a cosmonaut the choice was perfectly simple: take the chamber with a smile or you did not get to fly in space.

Gagarin survived several sessions without major incident, although he did confess afterwards that he found the experience 'uncanny, unnerving'. The psychologists told him to describe his thoughts and emotions at certain regular intervals timed by a clock inside the chamber. When he talked, his listeners did not always reply. He could not be sure whether they were ignoring him deliberately, or whether they had simply gone away to find some breakfast. Or maybe supper . . . Gagarin's clock gave no reliable clues to 'outside' time; nor could he count on dawn or dusk to guide his pattern of existence, because the chamber had no windows. The interior electric lighting came on when he tried to sleep or switched off abruptly when he was awake and busy with some task or other. Time lost all its meaning. 'It's easy for your mind to dwell too much on the past in such isolation,' he reported later, 'but I concentrated on the future. I shut my eyes and imagined myself in the Vostok, with the continents and oceans drifting beneath me.'[5]

The journalist Lydia Obukhova was allowed to witness one of these tests (though she was not permitted to publish what she had seen until some months after Gagarin's eventual flight in April 1961):

> Gagarin joked with himself in the chamber, and would speak through the microphone to whoever was on duty outside, even though he couldn't expect any answer . . . A few days went by. Outside the chamber, everyone knew that his isolation was to end that day, but Gagarin himself had no inkling of this. He started singing to himself about the few objects in the chamber with him. 'My electrodes . . . One electrode with a yellow wire . . . Another with a

red one.' The doctor explained, 'He has run out of stimuli in the chamber, so now he's looking for new ones, like a nomad in the desert singing about everything he sees.'[6]

Regardless of the mental torture inflicted by the psychologists, Gagarin never lost sight of the main prize: getting into space. He smiled, he charmed, he played the innocent farmboy to Titov's stern, poetry-spouting intellectual, but he tolerated all the tests with the same self-discipline and bravery.

When it came to Titov's turn in the chamber, he was one of the few cosmonauts to think more deeply about what the ordeal was really supposed to prove. It wasn't just a question of testing the body in various atmospheric pressures, or merely surviving the boredom. There were more subtle tests to pass, he was sure. 'They tell you there's no noise in the chamber, but that's rubbish. The air-conditioning system is working, the ventilator is on. They're all making noise, and you quickly get used to it. The most important thing is the isolation. Can you spend ten days on your own? . . . There's nobody looking through the keyhole, but you know you're being watched.'

The tins of food and the little cooking stove seemed to represent some sort of test. There was water in the chamber for drinking and for personal hygiene, but very little to spare. On the very first day of their sessions, some of the more impetuous cosmonauts had ripped open their food tins with great enthusiasm, eager to relieve the tedium with a snack. They had emptied the tins into the single saucepan provided, warmed the pan on the stove and eaten the food, only to find that there was no obvious means of cleaning the pan afterwards, and they still had many more mealtimes ahead of them. Titov says, 'I thought I'd be clever about it. I put my tins into a saucepan of water and heated them that way. Then you open the tins, eat the meals and throw the tins away. You don't have to wash anything afterwards.' Repeatedly boiled, the few cupfuls of water in Titov's pan were safe to reheat later. His water lasted, his pan stayed clean and his psychologists were happy. Or at least they were not unhappy, which was the most a cosmonaut could hope for at the end of his time in the chamber.

The doctors disapproved when Titov wanted to read in the chamber, but he got the better of them. He asked whether they could find a copy of *Yevgeny Onegin* for him to take inside with him. The answer was no, absolutely not. Such a form of entertainment was not allowed. Titov assured them he only wanted the book as a physical talisman, a foolish good-luck charm. 'I said I already knew it by heart. I wooed them. I persuaded them, and eventually they gave it to me. Of course I didn't know it!' So Titov contentedly frittered away his time in the chamber reading his book.

Film footage still survives of other sessions in the chamber, with Titov happily quoting Pushkin poems straight from memory, while the doctors observe him through a thick plate-glass window. He appears supremely confident, beaming with pride at his clever memory, his wide knowledge of literature. Unfortunately he was thinking too much like a bourgeois, assuming that his higher-than-average standard of education must be an advantage. In time, and to his great disappointment, Titov would discover that he had got it all wrong.

He was not so self-confident in the whirling centrifuge, a small capsule spun on the end of a long boom to simulate the g-loads of acceleration and deceleration: the traditional fairground ride for all pilots and astronauts the world over, loved to death by each and every one of them. For a pilot as proud as Titov, it was unnerving to be so much at the mercy of others. 'In an aircraft you can fly a high-g loop, and you control when you come out of it. The centrifuge was disgusting. The g-force is pressing you and pressing you, but you have no control. You just sit there like a guinea-pig.'

Gagarin did not like it either, despite his much-heralded talent for taking g-forces. His earlier qualification runs in an Air Force centrifuge had peaked at around seven g's. His MiG fighter had pulled nine, maybe ten on a vicious high-speed turn. Now the space-training centrifuge took him briefly up to twelve. 'My eyes wouldn't shut, breathing was a great effort, my face muscles were twisted, my heart rate speeded up and the blood in my veins felt as heavy as mercury.'[7]

Perhaps the most unpleasant training procedure was the 'oxygen starvation' experiment. The cosmonauts were locked in the isolation chamber while the air supply was pumped out, slowly but remorselessly. Gagarin had to endure this test without complaint if he was to be pronounced fit for a space mission. The doctors sealed him up, then watched his face on a television monitor while he wrote his name on a paper pad, over and over again. Journalist Lydia Obukhova watched this procedure, and archive footage taken by the technicians still survives. As the oxygen level in the chamber was reduced, Gagarin's writing became erratic, until eventually he was scribbling gibberish. Down went the air level. Gagarin dropped his pencil, dropped the pad, stared at nothing and suddenly blacked out. His threshold of consciousness must have been high enough for him to pass the test, or else he would have lost his place in the cosmonaut squad, but it is reasonable to assume that he shared the other cosmonauts' intense dislike of the procedure, because it made them look so foolish in front of the doctors. Senior space engineers shared this distaste and wondered what the doctors thought they could achieve by suffocating their test subjects. The only way a cosmonaut could lose his air supply in orbit was if his capsule sprang a leak, in which case he would close the visor on his space helmet and switch to a separate emergency air supply. If both the cabin and the spacesuit sprang leaks, then the cosmonaut was as good as dead anyway, but the chances of this double failure occurring were negligible. In space, either you had air to breathe or you did not. There were no half measures. The senior rocket engineers thought the 'oxygen starvation' experiment was pointless, but the doctors did not, and that was that.

For the rattled cosmonauts, it was almost a relief to move on to the parachuting exercises, where they could tease the doctors because most of them lacked the courage to follow the pilots out of the aircraft's doorway. Their instructor was Nikolai Konstantovich, an expert parachutist with a record-breaking freefall jump from a height of fifteen kilometres to his credit. Future space crewmen might have to eject and parachute to earth at similar altitudes, and Konstantovich's job was to demonstrate all the things that could go wrong up there – and how to get out of them. For example, there was the 'corkscrew' problem, when a misfiring ejection seat throws the pilot into a sickening spin, or the seat fires cleanly, but the craft it's coming out of is tumbling. Either way the pilot cannot pull his parachute safely, because the lines will twist around each other like the strands of a rope, and the silk canopy will not unfurl. The pilot must stabilize his fall and give his 'chute a chance to open normally. Konstantovich taught his pupils deliberately to sabotage their own jumps, then to regain control, preferably some time before they hit the ground. 'That's very unpleasant,' Gagarin recalled. 'Your body starts spinning at great speed. Your head feels like lead, and there's a sharp, cutting pain in your eyes. Your body is drained of strength and you lose all sense of direction.'[8]

At least the parachuting felt more like proper space training. Once Gagarin and his comrades were in a plane and doing something useful, they felt better. More in control. For another group of victims, however, there was no such easy escape from the doctors' needles, gas tanks and isolation chambers.

Quite apart from the cosmonauts, another group of 'testers' was put through a similar series of medical procedures – similar but worse. These young men were selected from a slightly lower rung of the aviation-academic ladder. They were not necessarily fighter pilots or first-rate theoreticians. They were just averagely bright, fit young military men. During recruitment they were never asked outright if they would like to fly in space. They were offered a chance to 'participate', which was a subtle but important shift of emphasis.

The testers' job was to find out just how much a human body could take. Then the cosmonauts, who were somewhat less expendable, could be pushed to those limits but no further. Unlike the cosmonauts, testers were not professionally acknowledged, and were paid only according to their previous military occupations as soldiers, technicians or mechanics. They were seduced with great care by their recruiters into a feeling of privilege and self-worth, but in truth their status was barely better than that of disposable laboratory rats. When they received injuries – and they did receive injuries – there were no special arrangements to compensate them or their families, because the authorities were unwilling to acknowledge any of their work in public. Even today, long after *glasnost*, the Russian space authorities do not like to discuss the testers' contributions to the early space effort, nor to the development of high-performance jet fighters, parachutes, ejection systems and flight suits for the Air Force. In all, approximately 1,200 testers were involved in various programmes over three decades.

They were all military volunteers, good soldiers, who did not like to surrender, to fail in front of their comrades. They were 'free' to abort any test-run halfway through if the discomfort became too great, but very few wanted to quit. Like the cosmonauts, who all dreamed of getting the first flight into space – of going higher, flying faster – the testers had their own strange pinnacle of glory to climb. Who could take the highest air pressures? The lowest? Who could survive the catapult sled's most rapid accelerations? The bone-jarring crash-stops? Who spent the longest time aboard the centrifuge, and how many g's could they take? Who among them was the toughest, the strongest, the bravest?

Sergei Nefyodov, a veteran from those days, recalls with a bitter smile, 'At first we didn't know what kind of tests we were in for, but it soon became seriously clear. They said they'd try us out in a "soft" landing experiment. That made us laugh! The tester had to throw himself out of a seat at some height – not great, but high enough. It was the level that might actually occur from a real landing of a spacecraft. There were some traumas as a result. The most serious was when something broke, or the system didn't work properly. Some lads couldn't get up again after the test.'

Nefyodov still boasts today that the testers took turns in a centrifuge that would have wiped out the delicate little cosmonauts. 'I achieved seven minutes at ten g's. The cosmonauts only had to endure two or three minutes at seven g's, and twenty seconds at twelve g's. My colleague Viktor Kostin often took twenty-seven g's for a very short time, by taking severe shocks in the catapult sled. Those were very brief, measured in microseconds, but once he went up to forty g's for a fraction of a second. I'd like to make it clear that we weren't chasing record g's just for the sake of it. We wanted to see how long a person could endure. Though we didn't use the word "record", because we couldn't claim any sporting achievements.'

This was the great frustration – the testers could not tell anyone how tough they were, because their jobs were extremely secret. They knew perfectly well that they were being given a much harder time than the cosmonauts. Sometimes they would look across at the physicians, with their dials and monitoring equipment, and they could only wonder. Nefyodov says, 'On the one hand, they [the doctors] were representatives of very humane professions, but on the other hand, the g-forces would climb, and the technicians would ask: should they stop the tests? It seems the subject can't take any more, he's going red, his heart is galloping, his sweat is flowing, but the doctors don't stop the test . . .

'It was dangerous for the testers. One of the senior academics for the test programme, Sergei Molydin, he said – and I can quote him – "we experimented on dogs, and fifty per cent of them survived. As you know, a person is stronger than a dog." Well, that's a joke! The consequences of our tests couldn't be predicted. Even if a person survived, he might become an invalid in later life, with damage to the lungs, the heart or other internal organs. Of course we could refuse a test at any time, but the unwritten rule was not to refuse. If you turned down a test, this would only happen once, because after that you weren't in the team any more.'

Nefyodov says that half of all the testers he worked with in the 1960s have not survived into the 1990s, but he isn't bitter about his career. Far from it. He takes great pride in his contribution to the space effort. 'The only tragic side is that our profession never existed. It was a close secret, so we had no social protection from the State, and no one ever investigated the long-term health of the testers. Today our old friends, our colleagues, are beginning to die.' He remembers a particularly nasty experiment to simulate the failure of a spacecraft's air-cleansing system. 'If carbon dioxide builds up in the air to three per cent in a submarine, for instance, that's enough to declare an emergency condition. With a colleague in a test chamber [at the Institute for Medical and Biological Problems] I went over to three and a half, four, five per cent. Honestly, you have nothing to breathe, your face turns an unpleasant colour, the lips turn blue, your brain won't work, you have a hideous headache and your strength fails. My partner and I had nosebleeds, but we worked for the set time. I always remember saying to him, "Just a half-hour more, and it will pass." Then half an hour later, "Just one more half-hour." I had to pep him up in some way.'

Yevgeny Kiryushin vividly remembers the altered state of consciousness he experienced in a centrifuge, with his body pushed beyond any normal definition of stress. 'Suddenly there was a light, it was very interesting, dark at first, then yellow, lilac, crossing some sort of emptiness, and then you just forget any sensation in yourself. You just have the impression that you're a brain, a hand, an eye. The oppressive weight is all in the seat, and suddenly, above your body is you. You're completely weightless, as if having a look at yourself from above. That's the transforming moment. All of your real achievements happen in those few minutes. But the experiments are frightening, without exception.'

Nefyodov remembers meeting Yuri Gagarin personally for the first time on January 2, 1968, when the First Cosmonaut came to visit the medical experimental facility and share a New Year's celebration with the testers. 'At that time I had just started working on explosive decompression, which interested me greatly. He kept asking me, "What's it like? Aren't you frightened? Have you simulated an atmospheric drop to fifty kilometres' pressure?" It was great to talk to him about all this. But suddenly he asked me why I was looking so sad. I felt calm, but to him it appeared as if I was sad. He gave me a warm hug and said to me, "Sergei, everything is in your hands. You must have an unsurpassable desire."'

An unsurpassable desire. The cosmonauts had it, and the world sang their praises. The testers had it, but they could not speak of it – except to Gagarin, who took the time to understand when a young lad volunteering for an insanely dangerous run of explosive decompressions tried, for a few moments, inarticulately, to share the reasons why he was doing it.

In fact, there seemed to be no shortage of people willing to volunteer for the most dangerous kind of work. Vladimir Yazdovsky, a senior manager closely involved with all aspects of the early Soviet space effort, recalls, 'After the dog Laika's flight in the second Sputnik, there were nearly 3,500 applications to the Academy of Sciences from people in prison, from abroad and from many organizations, saying, "You don't have to save me, just send me into space." Of course we couldn't reply to everyone, and as long as we didn't know how to bring someone back at the end of a flight, we weren't going to send anyone up.'

3

THE CHIEF DESIGNER

There was one man who, more than any other, dominated Gagarin's life from now on, at first behind the scenes, and later as a friend and powerful protector. This man was not a cosmonaut, though he had learned to fly long ago in his youth. He was a shadowy figure somewhere in the very highest echelons of space management. Most people within the Soviet aerospace industry called him 'The King' or the 'Boss of Bosses', or affectionately used his first two initials, 'S. P.' His full name was never spoken in public because the authorities had declared his identity an absolute State secret. In the many press and radio reports of Soviet rocket achievements broadcast over the years, he was referred to only as the 'Chief Designer'.

Born in 1907 in the Ukraine and educated in Moscow, Sergei Pavlovich Korolev began his career as an aircraft designer in 1930, before developing a fascination with rockets. At first he saw them as a useful power source for his aircraft, but by the late 1930s he recognized that rockets had a special potential as vehicles in their own right.[1]

Pre-war military strategists showed a keen interest in the work performed by the early rocket pioneers. Marshal Mikhail Tukhachevsky sponsored a new research centre, the Gas Dynamics Laboratory, hidden away behind the heavy ramparts of the Petropavlovskaya Fortress in St Petersburg (scorchmarks in the masonry can still be seen), while another laboratory in central Moscow, the Reaction Propulsion Laboratory, worked on similar problems. From these parallel efforts Korolev's rival, Valentin Glushko, emerged as the most promising designer of rocket thrust chambers and fuel pumps, while Korolev himself thought in broader terms about how to combine the engines with fuel tanks, guidance equipment and a payload, to create rocket vehicles that could perform some kind of useful work – delivering bombs, making weather measurements in the upper atmosphere and, one day, exploring space.

Marshal Tukhachevsky was primarily interested in winged rocket bombs and other useful armaments for the Red Army. In 1933 he began a major consolidation of the various rocket programmes into a unified military programme. Unfortunately Stalin was terrified of intelligent soldiers, and by 1938 he had initiated a wide-ranging purge of the officer class as part of his general regime of terror throughout all levels of Soviet society. Tukhachevsky was arrested on June 11, and was shot dead that same night.[2] Immediately all the rocket engineers he had assembled and sponsored came under suspicion of harbouring anti-Stalinist sentiments and were arrested. Korolev was dragged away on June 27 and was sentenced to ten years' hard labour in Siberia: essentially a death sentence.

In June 1941 the invading Nazis achieved devastating victories against a thoroughly unprepared Red Army. Stalin soon had cause to regret his earlier purge of the officer corps. His remaining commanders were almost exclusively talentless toadies with little experience of warfare or military strategy. Sergei Korolev and many other vital engineering personnel were released from their prison camps so that they could work in aircraft and weapons factories – still under guard. Towards the end of the war Korolev was freed from captivity, with his reputation partially reinstated, and in September 1945 he was allowed to venture into the crumbling German heartland in search of any remnants of Wernher von Braun's brilliant but infamous V-2 rocket programme that the American forces had not already taken away. Then, throughout the 1950s, with incredible energy and determination, Korolev developed an increasingly sophisticated range of rockets and missiles, while Glushko designed some of the most effective propulsion engines that the world had ever seen.

For a man who had survived a Siberian labour camp, mere administrative battles with competitors or unhelpful officials at the Kremlin must have seemed relatively easy, particularly under the far less oppressive post-Stalinist regime of Nikita Khrushchev throughout the late 1950s and early 1960s. Korolev was intensely driven, and he established a network of influence far more complex and subtle than anything his rivals in the aerospace sector could muster. By 1956 he was securely in control of his own industrial empire, the heart of which was a secret factory facility in Kaliningrad, just north-east of Moscow, known only as the Special Design Bureau-1 (OKB-1). Here Korolev was the absolute ruler, although he was answerable at Kremlin level to the Defence Ministry under Marshal Ustinov, and also to the vaguely named Ministry of General Machine Building. (In this context 'general' meant nothing of the kind; it was a cover word for rockets and satellites.)

In 1961 the Moscow journalist Olga Apencheko described the effect that Korolev seemed to have on those around him, as he strode through the corridors and shop floors of OKB-1, although she carefully avoided naming either him or his factory. As per regulations, she referred to him only as the 'Chief Designer':

> A dark-complexioned, rather severe-looking man with mas-
> sive features, the Chief Designer of the spaceship had
> something more in him than he cared to show. I heard
> a busy rustle around me whenever he appeared in a room
> or work area. It was difficult to say what this whisper
> expressed – awe, respect, a mixture of both. When he
> entered a workshop everything changed somehow. The
> movements of the technicians became more collected and
> precise, and it seemed that even the hum of the machines
> assumed a new overtone, intense and rhythmic. This man's
> energy stepped up the motion of shafts and cogs.[3]

Yuri Mazzhorin, one of Korolev's senior experts on guidance trajectories, says that he was 'a great man, an extraordinary person. You could talk to him about simple as well as complicated things. You'd think his time in prison would have broken his spirit, but on the contrary, when I first met him in Germany when we were investigating the V-2 weapons, he was a king, a strong-willed purposeful person who knew exactly what he wanted. By the way, he was very strict, very demanding, and he swore at you, but he never insulted you. He would always listen to what you had to say. The truth is, everybody loved him.'

Almost everybody . . . Valentin Glushko, an equally driven personality, operated out of his own specialist design bureau. As long as his engines were fitted into Korolev's rockets, the two men avoided outright confrontation, but these two giants of Soviet rocketry did not get along. The tension between them undoubtedly dated back to the summer of 1938 when, for some reason, Glushko was punished with eight months of relatively mild 'house arrest' while Korolev was sent to a prison camp. Presumably Glushko betrayed most of his colleagues, while Korolev kept a costly silence. Mikhail Yangel was another rival, developing missiles strictly for military use from his bureau in Dnepropetrovsk in the Ukraine; while the fourth major figure in Soviet rocket development, Vladimir Chelomei, had the presence of mind to hire Nikita Khrushchev's son Sergei as an engineer.

By far the most serious challenge to Korolev's autonomy came from high-ranking military officers based in and around the Kremlin and the Ministry of Defence, who were concerned that his space projects were blocking the development of necessary weapons systems. He outflanked them by creating a dual-purpose missile-space launcher, and then proving that his design for a manned spaceship could be adapted as an unmanned spy satellite. By satisfying significant military goals in tandem with his own, Korolev out-manoeuvred both Yangel and Chelomei, and maintained a firm grip on most of the important Soviet space programmes up until his death in 1966. His genius, unmatched by the engineers at NASA, was to standardize many of his principal spacecraft components so that a dazzling succession of manned and unmanned vehicles could be assembled around similar hardware.

American space analyst Andy Aldrin (son of Apollo 11 astronaut Buzz Aldrin) explains Korolev's cunning. 'The people who were running the military missile programmes had been with him since the war. To a large extent they owed their careers to him, so they were unwilling to take him on frontally. At the same time, the military didn't really understand his technology, and they implicitly trusted him. So when Korolev said, "Spy satellites won't work yet, we have to [lay the groundwork and] develop manned capsules first," they had little choice but to take him at his word. You could say, he conned them . . . He really understood how to work the political system.'

Academician Mtislav Keldysh was one of the Chief Designer's most powerful allies, consistently supportive of new space missions and scientific experiments in orbit. He was an expert in the mathematics of missile and rocket trajectories, and had established a power base in Moscow centred on his huge custom-built computing facility. Like Korolev, he possessed indispensable skills, with political cunning to match. While Korolev built the rockets, Keldysh plotted the routes they would fly.

The most irritating problem for Korolev was that he still had to rely on the Red Army's unintelligent cooperation during space launches, because his work with rockets and missiles was so intimately linked with military areas. He could subvert military equipment to his own ends, but he could not make the jealous generals vanish. However, if one of them blocked him, he was quite unafraid to treat the man as an inferior. A senior engineer in Korolev's bureau, Oleg Ivanovsky, recalls, 'On one occasion a very high-ranking commander refused access to an important radio communications link during a space flight. Korolev spoke to him on an open phone line and shouted, "You don't know how to do your job! Give me the link, or I'll have you demoted to sergeant!" We were amazed that he could be so insolent to a superior.'

First Secretary Nikita Khrushchev and his colleagues in the Politburo could be very supportive towards Korolev when it suited them, although they were not particularly concerned with the subtleties of space hardware. Rocket technology fascinated them more for its glamour and potential political impact than for its hard engineering details. When Korolev's primary missile and rocket development programme began in 1955, he asked senior members of the Politburo to inspect his work, as Khrushchev recalls in his memoirs:

> Korolev came to a Politburo meeting to report on his work. I don't want to exaggerate but I'd say we gawked at what he had to show us, as if we were a bunch of sheep seeing a new gate for the first time. Korolev took us on a tour of the launching pad and tried to explain to us how the rocket worked. We didn't believe it could fly. We were like peasants in a market place, walking around the rocket, touching it, tapping it to see if it was sturdy enough.[4]

Korolev's colleague Sergei Belotserkovsky (responsible for the cosmonauts' academic studies) sums up the Politburo's stance on space affairs. 'The top people's attitude to Korolev was purely that of consumers. For as long as he was indispensable, for as long as they needed him to develop missiles as a shield for the Motherland, he was allowed to do whatever was necessary, but the manned space research had to follow on the back of the military work. The point is that Korolev launched his cosmonauts on the very same missiles.'

His principal workhorse was the dual-purpose R-7 missile-space launcher, or 'Semyorka' ('Little Seven'), as it was affectionately known by the men who built it or flew on it. Fuelled with liquid oxygen and kerosene, and incorporating four drop-away side-slung boosters, this was the world's first operational inter-continental ballistic missile (ICBM). Each stage or 'block' of the vehicle was fitted with one of Glushko's four-chamber engines. It has to be said that Glushko's engines were superb – in fact, they are still in use today in the upgraded R-7 rockets that carry modern Soyuz capsules to the orbiting Mir space station. Glushko's innovation was to design compact fuel pumps and pipework to service four combustion chambers simultaneously. The apparent thrust of twenty separate engines on the R-7 is, in fact, delivered by five.

The first two launches of the R-7 failed, but on August 3, 1957, it flew successfully in a simulated ICBM trajectory, then began its career as a space launcher just two months later, on October 4, by launching 'Sputnik', the world's first artificial satellite.

Andy Aldrin is full of admiration for the speed with which Korolev could conjure up his space triumphs. 'He and his merry band of rocket engineers tried to go off on a vacation after Sputnik, and they'd rested for about two days when Korolev got a call from Khrushchev. "Comrade, we need you to come to the Kremlin." Of course he went, and he sat down with the Soviet leadership, and they said, "In a month we have the Fortieth Anniversary of the glorious October Socialist Revolution. We want you to put up another satellite that will do something important." They proposed a satellite that could broadcast the "Communist Internationale" from space, but Korolev had another idea. He wanted to put a live animal in the satellite, so that he could lay the groundwork for an eventual manned mission; and within a month, from scratch, he and his people completed the spacecraft and launched it.'

Sputnik II went up on November 3, carrying the dog Laika. This was a clear indication of where the Soviet space effort was heading. 'Americans were shocked by Sputnik, and then Laika. This dark, mysterious, backward country on the other side of the world, that was considered to be thoroughly nasty, had jumped ahead of them.'

A small American rocket at last carried their first satellite, 'Explorer I', into orbit on January 31, 1958. Khrushchev disparaged it as 'a grapefruit' because it weighed only 14 kg against Sputnik's 80 kg and Sputnik II's 500 kg – although Explorer I immediately made one of the most important scientific discoveries of the twentieth century when Dr Van Allen's simple instruments detected radiation belts around the earth.[5]

When it came to selecting candidates for the manned space programme in the autumn of 1959, Korolev looked at all the most promising personal files, but it was not until June 18, 1960 that he summoned the twenty successful applicants to OKB-1 in Kaliningrad to see an actual spaceship. (The hardware had not been anywhere near completion until that time.) Alexei Leonov remembers the Chief Designer introducing himself with a little speech designed to put the cosmonauts at ease. 'He said, "What we're doing is really the easiest thing in the world. We invent something, find the right people to build it properly, and place lots of orders for components with the best and most experienced factories all around the country. When they at last deliver what we've ordered, all we have to do is put the pieces together. It's not very complicated." Of course we knew there was more than this to the building of a spaceship.' But the cosmonauts were touched by Korolev's warmth and friendliness. 'My little eagles' he called them.

According to Leonov, Yuri Gagarin made a good impression in Korolev's office that day, listening intently and asking pertinent questions about space and rockets. In this formal semi-military context – young recruits being introduced to a superior for the first time – Gagarin's curiosity might easily have been misread for impertinence, but the Chief Designer was pleased that any of the cosmonauts should ask direct questions. Leonov remembers, 'He told Yuri to stand up, and he said, "Tell me, my little eagle, about your life and your family." For ten or twenty minutes it was as if Korolev forgot about the rest of us, and I think he liked Yuri immediately.'

Gherman Titov, a somewhat prideful man, was not cowed by the Chief Designer's reputation and authoritarian demeanour. 'What did I know, a young lieutenant with eyes full of courage and scarcely a single sensible thought in my head,' he admits ruefully. Over the coming months and years his relationship with Korolev never really developed into genuine warmth. 'Probably, two lions couldn't exist in the same cage. I don't want to say I was ever the same calibre of lion as Korolev, but we did have quite a difficult relationship.'

It was Yuri Gagarin and Alexei Leonov who emerged as firm favourites to be taken under the Chief Designer's wing, although he would prove fiercely loyal and protective towards all the cosmonauts who flew for him and put their trust in his rockets and capsules.

After this first introduction Korolev escorted the cosmonauts into the heart of OKB-1. They went into the main construction area, while Korolev and one of his senior spacecraft designers, Oleg Ivanovsky, started to explain what they were seeing, but it was hard to take things in. There were a dozen spacecraft, lined up neatly, their positions in the rank depending on their current state of construction, from bare shell at one end to near-completion at the other. Archive footage still conveys the extraordinary scene. Each ship consisted of a silvery sphere mounted on top of a conical base covered in wires and pipes, with another reversed cone beneath it, clad in delicately grooved metal vanes. The double-cone section was a detachable equipment module, and the vanes on the lower cone were radiators. The big spheres (everybody called them 'balls') were cabins for the crew.[6] The machines had no aerodynamics, no control surfaces or any obvious means of propulsion; no proper landing gear, even. They could not stand on the floor properly, but had to be supported inside metal frames to keep them upright, like unstable buildings propped up with scaffolding. 'It was something we couldn't grasp at all,' says Titov, 'It was completely incomprehensible to us – a ball without wings, without anything. It wasn't easy for a pilot to understand. Of course, as pilots, we'd never come across anything to compare it with.'

These were 'Vostok' space capsules.

There were bundles of electrical wires running from various test boxes and conduits in the factory that snaked across the floor, sprouted from the walls, dropped from the ceiling. Every last one of them plugged, like sinister roots, into the space machines, testing and probing, powering them up, shutting them down, as the white-coated engineers ran their countless tests.

Oleg Ivanovsky, who had a reputation for being long-winded, lectured the cosmonauts in mind-numbing detail about the ship's components. 'They dragged a few words out of me, as they say.' All the while, he scanned the faces of the twenty would-be cosmonauts before him; twenty young men, twenty names, twenty strangers. Korolev knew them by now, of course, but Ivanovsky had not met any of them until this moment. 'They all gazed at the vessels with great curiosity, as this was the first time they'd seen any space technology. I knew they were all pilots, familiar with aviation, but one has to say honestly: which of them could get used to this new apparatus?'

Korolev cut short Ivanovsky's lecture and explained Vostok's flight characteristics in terms that his audience of MiG-trained pilots could more easily understand, but he warned his listeners, 'There's a lot you have to learn. We can't tell you everything in one day. We'll prepare special classes, so that you can learn the system thoroughly. You'll attend lectures, and then we will set you some exams.'

One of them, a handsome lad with an irresistible grin, asked Korolev a question. 'Sergei Pavlovich, will you be marking us?'

'Yes, and we'll throw you out!' Korolev barked at him. 'Stop smiling! What are you smiling at, Yuri Alexeyevich!'

Korolev glared at the boy to see how he would react, perhaps deliberately trying to undo the mood he had created back in his office. Gagarin forced the muscles of his face into a sober expression, but he was not at all frightened. He stayed perfectly calm, and almost certainly this was the response which Korolev was looking for. Ivanovsky had discovered this from personal experience. Just a few weeks previously Korolev had thrown a fit and sacked him on the spot. He often sacked people, then reinstated them the next morning. It was his way of letting off steam. On this occasion he yelled across the assembly floor, 'You no longer work for me, and I'm putting a de-merit in your record!' Ivanovsky shouted back, 'You can't do that, because you just fired me. I don't work for you any more!' Korolev shouted at him again, but in a short while the incident was forgotten. The Chief Designer admired people who stood up to him – people who played straight, and did not hide important problems under the table simply in order to protect their jobs. His relationship with Ivanovsky was very trusting after that. And now, here was this cocksure farmboy from the Smolensk Region . . .

Suddenly Korolev invited the cosmonauts to take a closer look at one of the Vostoks, a version to be used in ground tests, but fitted just the same with most of the equipment used during an actual flight, including the ejection seat and control panels. Alexei Leonov remembers Korolev telling them to take off their boots (to preserve the ship's cleanliness), then go up a ladder and climb through the open hatch of the ball. Without a moment's hesitation Gagarin stepped forward. 'With your permission, Sergei Pavlovich?' He sloughed off his boots and clambered up.

One of the other cosmonauts, Valery Bykovsky, insists that Gagarin was not actually told to remove his boots. After all, a pilot would not expect to take off his footwear to sit in the cockpit of a new MiG, so why now? 'That's how they take off their shoes in Russian villages when they go into a house, as a sign of respect,' Bykovsky thought to himself. He was sure that Gagarin became 'the one' from that moment.[7]

Gagarin paid no attention to his companions, all busily removing their shoes on the floor behind him. He was much too fascinated by the spacecraft cabin. It was swathed throughout in a light tan-coloured rubber foam. This cladding disguised the sphere's real guts: the endless pipework and electrical distribution systems. It would be some weeks yet before any of the cosmonauts was introduced in more detail to these shrouded mysteries. For now, Gagarin's quick inspection took in only the most obvious items. He would have found the interior much less cluttered than the cockpit of his MiG. Certainly there were fewer dials and instruments. The ejection seat, which he now reclined on, took up much of the space. This must have seemed reassuringly familiar, except that he had to lie on his back rather than sit upright. Mounted on the wall directly above his face was a simple panel with a few switches, some status indicator lights, a chronometer and a little globe representing the earth. To the casual observer it might have looked like a child's educational toy, but in the months to come, Gagarin and his colleagues would learn about the hidden gyroscopes and accelerometers that fed their data into the globe, allowing it to swivel in precise lockstep with Vostok's orbit, relaying to the pilot his position over the real planet earth. Other indicators on the panel counted off the orbits, and gave readings of the ship's internal temperature, pressure, carbon dioxide, oxygen supply and radiation levels. The cosmonauts would discover that these displays were not necessarily intended directly for their benefit.

They were supposed to scan the dials at intervals, then report various measurements to the ground by radio link. Others would decide what the readings meant, and what should be done with the ship.

To his left, Gagarin would have seen another small panel with four rows of toggle switches. Leonov says that he fiddled with some of them, understanding straight away that he could only get to these left-hand controls by reaching across with his right arm. The natural armchair-style arrangement of an aircraft cockpit obviously did not apply here, although his right hand in repose settled comfortably on a small lever, the nearest equivalent any Vostok crewman would find to the control joystick in a fighter plane. Higher up, on the right, there was a radio transceiver. The only other piece of equipment obviously accessible to him was a food locker. It did not seem to matter which hand one used to reach that.

Below and slightly forward of his feet, Gagarin would have seen a round porthole with elaborate calibrated markings. The 'Vzor' was an optical orientation device consisting of mirrors and lenses. Through this, the curvature of the earth's horizon would appear greatly exaggerated. When Vostok was aligned at a particular angle relative to the ground, the pilot would see a brilliant circle of horizon all around the Vzor's outer edge. This would indicate that the craft was properly positioned for its re-entry burn. The effect was rather akin to a distorting mirror at a fairground, adapted by ingenious but under-equipped engineers into a precision tool of space navigation.

The layout of this Vostok test cabin was not exactly the same as Gagarin would find aboard the actual flight version. In time he would encounter more equipment: for instance, an intrusive television camera that pointed directly at his face, and a bright lamp that shone uncomfortably into his eyes so that his every expression could be recorded for the doctors. And, on the left-hand switch panel, a numeric keypad – two rows of three numbers, six digits in all – whose purpose was not immediately clear to him or any of the cosmonauts who clambered into the cabin after him.

They called Gagarin out after a few minutes. The other cosmonauts took their turns in the cabin, while Korolev and Ivanovsky leaned into the hatch to show them the controls.

Afterwards, while they were leaving the construction hall, the cosmonauts chattered eagerly about Vostok and who would be first to fly it. Alexei Leonov remembers putting his arm around Gagarin and saying, 'Believe me, today could've been very important for you. I know you can be the first to go.' Several in the group chimed in their agreement. Gagarin had made an impression back there.

Meanwhile Valya was beginning to understand the pitfalls of being a cosmonaut's wife. As she told journalist Yaroslav Golovanov in 1978, 'Yuri would often come home late, and he frequently went on trips to do with his work. He wasn't very communicative about what he did, and if I ever showed any curiosity he would dismiss it with a joke. I know he wasn't allowed to talk about these things even with his family, but it sometimes seemed to me that his work at Star City was taking him away from me more and more. I tried to make it seem that I hadn't noticed, but from time to time I would be overcome by a strange anxiety.'[8]

One day towards the end of 1960 Gagarin brought some of his cosmonaut friends home, and Valya, coming in from work at the Star City clinic, overheard them whispering. 'It'll be soon now. Either Yuri or Gherman.'

4

PREPARATION

When Korolev and his colleagues brought home captured German V-2s and began to fire them, they built a small testing station 180 kilometres east of Volgograd (then called Stalingrad) near a small town called Kapustin Yar. In January 1957 work began on a much larger and more permanent base at Plesetsk, on the Arctic Circle, because trans-polar trajectories offered the shortest ballistic routes into the North American continent. Plesetsk became the principal base for Soviet intercontinental nuclear missiles – though the 'missile gap' which John F. Kennedy spoke of so effectively in his 1960 election campaign was largely a myth. Kennedy presented a frightening vision of vast numbers of Soviet missiles aimed at the US, and urged the building of an adequate counterstrike force to reduce the supposed 'gap' between Russian and US capability. In fact, at that time Plesetsk could handle no more than four of Korolev's R-7 rockets at a time, and it is unlikely that they could have been launched all at once. The missile gap was, if anything, greatly in America's favour.

The Soviets' most famous launch station was built as close to the Equator as they could manage, so that the west–east rotation of the earth would impart extra energy to departing heavy rockets. On May 31, 1955, supervising engineer Vladimir Barmin and his men turned over the first clump of soil in one of the loneliest places on earth: a vast, utterly flat, barren steppe in the middle of the southern republic of Kazakhstan. The new complex was built around an old settlement called Tyura'tam, named by the nomad Kazakhs for the burial site of Genghis Khan's beloved son Tyura, although another translation was

'Arrow Burial Place', which was not considered appropriate for a rocket-launching station. The Soviets swept aside the old name and called the place 'Baikonur', which was actually a small town 370 kilometres to the north-east. This was a bid to confuse the Western Intelligence agencies about the base's location, although they discovered the truth as soon as the first R-7 ICBM prototype was successfully launched from Baikonur (after two failures) on August 3, 1957 and was monitored by radar stations in Turkey. Near the rocket base the Soviets founded a new city called Leninsk to house 100,000 Russian technicians, along with 30,000 soldiers to guard them.

Several metres of snow cover the steppe from October through to March, and blizzards are frequent. Only in April does the place become bearable, when the snow melts and the steppe comes into bloom for two or three weeks. As the flowers fade and the last of the meltwaters evaporate to leave shallow pools, the mosquitoes breed. Then, in the long summer, the earth hardens like brick, the heat is remorseless and sandstorms are a constant hazard for people and machines alike.

At first glance the engineers working on the Baikonur complex in 1955 could have been mistaken for political prisoners. They lived in tents, by turns freezing and sweltering, and their equipment was so inadequate that they had to start their work using just shovels and spades. Their first task was to run a triangular spur from the Moscow–Tashkent railway (which itself followed an ancient nomadic caravan route). While NASA supplied its launch centre in Florida with an endless succession of cargo planes, barges, helicopters and 16-wheel trucks cruising along smooth highways, the Soviets went into space by train. Only when the rail spur deep into the steppe was completed could proper construction machinery arrive at Baikonur.[1]

Within two years the construction workers completed an airport, a huge hangar bay where rockets could be assembled and checked under shelter, control blockhouses, and a support platform and flame trench for the base of the first launch tower. The 250-metre-long platform, supported on solid concrete pillars the height and size of apartment buildings, jutted out over the reinforced slope of an old mine working, like a giant balcony over a hillside. Rockets would be suspended by clamps with their engines pointing down through a large square hole in the platform, so that in the first moments of ignition the engines' blazing exhaust products would shoot through the hole and down onto the slope, to be deflected harmlessly away from the pad.

Other launch pads soon followed, and over the next decade Baikonur's various facilities sprawled across hundreds of square kilometres of the steppe. Until 1973 no American had ever seen this place, except as a vague pattern of rectangles, lines and shadows in high-altitude reconnaissance photos, taken at great risk by spy planes flying out of Turkey. In fact one of the worst embarrassments in the history of US reconnaissance occurred on May 1, 1960, when a U-2 aircraft was shot down over the Ural mountains. Its mission was to overfly Baikonur and photograph the launch pads. The pilot, Gary Powers, was captured and put on trial in Moscow, much to Khrushchev's glee. The US president, Dwight D. Eisenhower, in his last months of office, made empty protests about an unprovoked attack on 'an American weather research aircraft flying from a base in Turkey', which had 'inadvertently strayed off-course'.[2] Immediately Eisenhower banned any further U-2 overflights of Soviet territory.

In the wake of this humiliation, one of the most costly, secret and technologically sophisticated space efforts was born: the US spy-satellite programme, run largely by the CIA and the Department of Defense. Their projects came to be known as 'black', because nobody ever knew very much about them, despite budgets that matched, or even exceeded, the funds allocated to NASA's more visible space exploration projects.[3]

No longer a secret, the first R-7 launch pads at Baikonur are still in operation today. Stars painted onto the metal gantries denote the number of launches – one star for every fifty launches. One gantry is decorated with six stars ... This is the facility from which the world's first manned space mission was launched. Today it despatches Soyuz crew-ferries to the orbiting Russian space station Mir.

Baikonur's modern launch record is good, but the early years of the complex were dogged by failure. In particular, the six months prior to Vostok's first manned launch were extremely discouraging. On October 10, 1960, Korolev's robot probe Mars I reached a paltry 120 kilometres into the sky before falling back to earth like a damp squib. The base blocks of the R-7 booster fired according to plan, but the uppermost interplanetary stage, designed too hurriedly, failed to push the probe clear of the earth's gravity. Four days later, a second probe fell back in the same way. At the time Nikita Khrushchev was attending a United Nations conference in New York. He had looked forward to boasting about the Mars project, but an urgent coded telegram from Moscow changed his mind. He was most upset.

In mid-October a new prototype rocket, the R-16, was hoisted upright for launch at Baikonur. This was one of Mikhail Yangel's military machines, designed as a replacement for Korolev's R-7, which was proving somewhat frisky for space exploration and even worse as a strategic missile. If the Soviets were ever to deploy a truly credible force of ICBMs, they had to find a rocket capable of firing at much shorter notice. The R-7 was fine as it went, but it took at least five hours to fuel and prepare. The problem was its use of liquid oxygen, which was a highly efficient chemical when it was actually burning inside an engine, but would not keep for very long prior to launch. Inevitably, after a few hours it warmed up and turned from liquid to gas. The pressure in the tanks climbed towards bursting point and the accumulating gas had to be vented, then replaced with fresh supercold liquid. The longer an R-7 stood on the pad, the more the greedy creature needed replenishment.

The R-16 was designed to need far less preparation before launch, in keeping with the military's need for a fast-response missile. It could be fuelled and primed several days, or even weeks, before it was needed, with no loss of oxidizer, because Yangel had disowned supercold liquid oxygen and kerosene in favour of nitric acid and hydrazine. These chemicals could be stored for long periods inside the rocket at normal pressures and temperatures, without venting or leakage. The R-16 could be kept on permanent standby in a secret silo, ready to strike the Americans at a moment's notice. The only trouble was that its 'storable' fuels would not store. They were viciously corrosive and did just what they were not supposed to do – they leaked.

Thwarted by the Mars probe failures in October, Nikita Khrushchev remained quite determined to come up with a bold gesture for the United Nations conference, so he focused on Soviet military superiority. 'We're turning out missiles like sausages from a machine!' he crowed. On his return to Moscow he pressured his Chief of Missile Deployment, Marshal Mitrofan Nedelin, to come up with a tangible demonstration of strength. Khrushchev wanted no damp squibs this time. Nedelin flew to Baikonur straight away to supervise the début launch of Yangel's R-16 on October 23.

As zero-hour approached, the missile began to drip nitric acid from its base. What does a cosmodrome commander do when a fully fuelled rocket springs a leak? He drains its fuel away carefully and then pumps non-flammable nitrogen through the tanks to get rid of any lingering vapours. Next day he might send in a couple of brave technicians in heavy fire-suits to 'safe' the rocket, so that it can be taken down and checked. Instead, Nedelin sent dozens of ground staff to the pad straight away, to see if they could tighten up some valves, stop the leaks, and get the R-16 up in the air. His instructions seemed so insane that the crews were at a loss how to proceed. In the firing blockhouse, the proper thing to do was to reset all the electronic sequencers and disarm them, before they could send any further ignition signals to the rocket. Nedelin ordered the firing sequences to be revised and delayed, but not cancelled. Somehow, a wrong command was transmitted to the R-16's upper stage. Its engine fired, straight away burning a hole in the top of the stage beneath it. This lower stage exploded, instantly killing everyone on the gantry. With nothing to support it, the upper stage then crashed to the ground, spilling fuel and flame. The new tarmac aprons and roadways around the gantry melted in the heat, then caught fire. Ground staff fleeing for their lives were trapped in the viscous tar as it burned all around them. The conflagration spread for thousands of metres, a wave of fire engulfing everything and everyone in its path. More than 190 people were killed, including Nedelin, perched on his chair near the gantry, as a wall of blazing chemicals swept towards him.[4]

For thirty years the West knew little of this, although it was apparent from various Intelligence reports that something had gone awry. In particular, an American Discoverer spy satellite photographed Baikonur the day before and the CIA noted with interest the stacking of a new missile. On October 24, the Discoverer in its predetermined orbit overflew the site once again and recorded no gantry and no rocket, just a very large dark smudge despoiling the landscape. The rocket had exploded, but so what? American rockets also blew up from time to time. One had to expect the occasional bad day. The scale of the disaster was not immediately apparent because all news of it was suppressed. All of Soviet Russia was saddened to hear (eventually) that Marshal Nedelin and several other senior missile officers had been killed in an 'aircraft accident'. Of course the absence of many familiar faces became obvious to thousands of space workers beyond Baikonur, but such unpleasant and difficult matters could be discussed only in private. The sudden disappearance of dozens of young military technicians from Yangel's squad – most of them just nineteen, twenty, twenty-one years old – was not so immediately apparent, except to their mothers.

Gagarin and his fellow cosmonauts were told that a prototype missile – not one of Sergei Pavlovich's 'Little Sevens' – had blown up and several technicians had been injured. No doubt they knew better, but for the time being they remained closeted from the worst of the horror in their training compound at Star City. In fact the explosion did not greatly delay preparations for Vostok. Surviving ground crews at Baikonur were able to continue their work. The pads, fuelling pipes and blockhouses assigned to the manned mission were not damaged, and few of Korolev's most important technicians had been actively involved with the R-16.

Then, less than three weeks before the first manned flight, one of the cosmonauts was killed. Valentin Bondarenko was the baby of the group, a fresh-faced lad of twenty-four years. When his turn came to go into the isolation chamber, he handled his assignment very well. His was a fairly long session (fifteen days), to see how he made out. On March 23 he prepared to exit the chamber. They were running a 'high-altitude' regime, and the chamber had to be brought up to normal pressure very slowly, or Bondarenko would suffer from the 'bends'. There was another half-hour to go before the supervising technicians could equalize the pressures and open the hatch. Bondarenko stretched, climbed out of his itchy woollen outer garment and peeled the medical sensor pads from his torso and upper arms with evident relief. He cleaned his irritated skin with pads of cotton daubed in alcohol. Perhaps he tossed the pads aside a little carelessly. One of them landed on the hotplate of the little cooking stove and caught alight. In the confined, oxygen-rich environment of the chamber, the fire spread with terrifying rapidity.

They pulled him out, covered in burns and in great pain. 'It's my fault! I'm so sorry!' he cried. The doctors struggled for eight hours to save him, but his injuries were too extreme. The circumstances of his death were not made public until 1986.[5]

There was one aspect of space flight for which the cosmonauts at Star City had no practical means of preparing themselves in advance: weightlessness. Korolev and his advisors were not keen to allow their first manned spacecraft to drift in space for longer than a single orbit, because no one was sure that its passenger could survive an entire day without the normal sensation of gravity.

Weightlessness presented a tremendous psychological barrier for the early Soviet space programme. The only earth-bound opportunity to taste the sensation was in the 28-storey lift shaft at Moscow State University, one of the city's tallest buildings. There was a special cage that fell freely down the shaft and slammed into compressed-air buffers at the bottom of its drop. The cosmonauts might float freely of the cage's floor for two or three seconds at best. Korolev's guidance specialist Yuri Mazzhorin explains, 'It was our first dive into an ocean of uncertainty. We were afraid of everything. That's why Sergei Pavlovich was in favour of a gradual approach. For the first human space mission, one circle.

The next flight, twenty-four hours. The next, three days, to see how a person would survive.'

American astronauts at NASA flew long parabolic arcs aboard Boeing 707 jet planes. These craft were essentially cargo-carrying airliners, but with all the seats and storage crates stripped out, so that the interior cabin was a capacious free space. The astronauts could float free of the walls for perhaps two minutes at a time – more than enough to eliminate the sinister mystique of weight-lessness. The Russians never thought to use their cargo planes in this way, at least not in the early 1960s. Trainee cosmonauts experienced thirty seconds or so of near-weightlessness while jerking about in the back seat of a MiG-15 fighter aircraft flying a similar parabolic arc, but it was barely more useful than plunging down the lift shaft at Moscow State University. Titov recalls the MiG experience being uncomfortable and unsatisfactory, and so short that it was not much more than he was used to on ordinary combat training missions. 'When you're performing an advanced manoeuvre and not doing it well, you might get something similar, and all the dirt and dust on the cockpit's floor flies into your face. These short bursts aren't weightlessness as such. It's very different [in space] when you have to live in a weightless environment for long periods.' What's more, the MiG cockpits were so cramped that there was little chance to float about in any meaningful way.

The Soviets' dread of weightlessness remained unconquered, at least for now. Calculations were made to fire Vostok's braking rockets well within its first orbit, which would keep the period of weightlessness to a minimum. However, there was the remote possibility that the craft might be stuck in orbit for several more circuits, because of a failure in the retro-rockets. Andy Aldrin neatly sums up the risk faced by the world's first space traveller: 'In an orbital trajectory what happens is you go up and around the earth, and then the rocket that got you up there has to work again to slow you down and bring you back in. If that doesn't work, you end up with a man orbiting the earth for ever and essentially dying . . . Korolev's designers suggested a safer sub-orbital trajectory for the first manned attempt, but he made it clear he didn't want to beat the Americans by such a little margin. He wanted to beat them by a lot.'

Vostok's air supply (sixteen spherical gas tanks, alternating nitrogen and oxygen reservoirs wrapped in a 'necklace' around the join between the ball and the equipment module) would last for a maximum of ten days. The ship's orbit was deliberately designed to skim the outermost layers of the earth's atmosphere so that, in the event of a serious problem, natural atmospheric friction would slow the craft down within a few days. It was a gamble whether or not this would happen before the cosmonaut ran out of air, water and food.

With help from Academician Keldysh and his computers in Moscow, Mazzhorin calculated that Vostok's re-entry ball could be recovered safely at the end of its first orbit, so long as the braking rockets in the rear equipment module functioned without incident. But even if these rockets were in good shape, there was a chance that the braking manoeuvre might have to be delayed while Vostok's orientation was fine-tuned. In theory the retro-systems could be fired at any time, but there was no guarantee that they would bring the capsule down anywhere within Soviet-held territory.

Vostok's orbit was inclined to the Equator by sixty-five degrees. Each orbit, west to east, took ninety minutes. Meanwhile the earth rotated at its own steady pace beneath the ship, once every twenty-four hours. As a consequence the craft did not fly the same path over the ground each time. The mathematics of the situation were clear. The best opportunities for a good homecoming occurred one hour into the first orbit, or else a whole day later, halfway through orbit seventeen. Firing the retro-rockets during any other orbit would risk bringing the craft down into the sea, or on foreigners' land; in which case, the embarrassment could be severe. Secrets of technology might be revealed; corrupt capitalists might claim the glory for 'rescuing' a cosmonaut within their own borders.

Eventually the solutions to these potential propaganda problems were sealed within three envelopes, addressed to the official news agency TASS in Moscow. The various enclosed documents were prepared by Mazzhorin, doubling not just as a guidance mapper but also as a propaganda officer. He had such a detailed understanding of how and where the ball might come down, at the end of its flight, that it seemed appropriate for him to work out what measures should be taken if it actually did descend onto foreign soil. If this calamity occurred, then TASS would be instructed to tear open the appropriate envelope and broadcast its contents. Mazzhorin was also instructed to prepare for the very worst-case scenarios. If the capsule blew up in space, or the ball sprang a leak, then the press statements would have to be tailored accordingly to make the best of the situation. It seemed wise to consider all possibilities and prepare the various announcements in advance. 'We prepared three envelopes for TASS, with different announcements,' says Mazzhorin. 'Envelope number one in case of a full success. Number two for a forced landing over foreign territory. Number three for a catastrophe. People in the television and radio stations were waiting. When we saw the cosmonaut had made it into orbit, and we had data, altitude, inclination and orbital period, the Kremlin could order TASS to open envelope number one.'

But even this 'success' envelope was not easy to compile. 'When the capsule, suspended on parachutes, reached seven thousand metres, the cosmonaut was supposed to eject and come down under his own parachute. We weren't sure whether to include this part.'

The problem was simple. In the event of a good flight, the Soviets intended to claim the World Aviation Altitude Record, according to rules established by international agreement. Korolev read these rules with care, and noticed to his alarm that any pilot claiming such a record had to remain inside his vehicle all the way to touchdown. If a pilot bailed out before landing, the rules assumed that something must have gone wrong with the flight. In which case, no record. The alternative was not to eject Vostok's crewman, but Korolev was not sure that anyone could survive the re-entry ball's abrupt landing without injury. Gai Severin, the foremost Soviet designer of fighter-pilot equipment, had already designed an emergency ejection seat for the cabin, just in case the

R-7 launched badly and the cosmonaut had to get clear of some terrible explosion. If the same system was used to remove him at the end of his flight, there would be no need to worry if the ball came down rather too hard on the ground. Future re-entry modules would incorporate larger parachutes and a cluster of rockets in the base to soften the final impact. In later years, more powerful upper stages for the R-7 would allow larger and better-equipped ships to be hauled aloft. For now, the power-to-mass calculations for the system allowed Vostok no margin for luxuries. Soft-landing rockets for the ball were not an option, and its crewman had no choice but to eject.

Nikolai Kamanin instructed a sports official, Ivan Borisenko, to research the altitude-record regulations more deeply. By February 1961 the problem still had not been resolved. At this point, very late in the day, a strategic untruth seemed much more appealing than a major re-design of Vostok. Mazzhorin's first envelope for TASS, containing the 'successful' announcement, falsely implied that the cosmonaut landed in his ship. 'For a long time this legend was supported in all the official documents,' says Mazzhorin. 'Only in the *glasnost* era was the truth revealed to our people, and to the world.' The other envelopes must have told a different story. For instance, if Vostok had landed over non-Soviet territory, the use of an ejection seat would have been blatantly obvious to foreigners. Mazzhorin cannot remember the precise wording he used, and he regrets the subsequent loss of his envelopes. 'It's a pity we destroyed them. They'd have a historic value today.'

Even the simplest details in the documents for TASS presented a challenge. In the lead-up to the first manned flight it seemed natural to call the capsule 'Vostok-1', in the expectation that others would follow in its wake; but the capsule's principal designer, Oleg Ivanovsky, recalls, 'If we'd given it a number, then it would have suggested that a series was beginning. We didn't want anyone to know we were preparing other flights, so Vostok wasn't given a number.'

A fourth and very different kind of document was prepared for stowage in the Vostok capsule itself. The cosmonauts were not to know it, but even in the last few weeks before the first manned flight, arguments were raging about the extent of command and control that a space pilot should be allowed during his mission. Everything centred around the mysterious six-digit keypad on Vostok's left-side control panel.

So far, all space vehicles had been operated by on-board electronic systems linked by radio to control stations on the ground, which represented a difficult challenge in itself. What new problems might arise when a human pilot was included in the ship? The doctors worried that a solo cosmonaut might go mad up there, overcome by the spiritual and psychological separation from his companions on earth, while the security services worried that he might defect to the West, deliberately re-entering his craft over foreign territory at the end of his flight. By the autumn of 1960 the discussions about control underwent a bizarre shift of emphasis. The aim was no longer to give the pilot some dignified authority over his own vehicle, but to take it all away from him. Guidance of Vostok would be purely automatic, just as with all the unmanned ships. In an emergency, the crewman might be allowed to operate the controls for a while, but only if he could prove his sanity beforehand.

The engineers devised a six-digit keypad that would unlock the navigation systems from the computers and let the pilot steer his own ship, if manual control became necessary. He would be told the keypad combination only if the mission directors on the ground decided he was mentally fit for the job. With his accustomed logic, Sergei Korolev broke this plan down into its component parts and questioned the basic assumptions. Why would a pilot be given control of the ship? Presumably because the automatic systems had failed and he needed to take over. But if the ship started to tumble out of control, the radio link with earth might be interrupted just at the point when the pilot really needed to hear the secret code that would release his manual controls. The keypad idea seemed more dangerous than just leaving things be.

The doctors came up with a face-saving solution whereby the pilot could find out the code even if his radio went dead, as Vostok's co-designer Oleg Ivanovsky explains: 'They decided that if he reached for an envelope placed inside the cabin, ripped it open, took out the paper and read the number printed on it, then pressed the keypad, this careful sequence of actions would prove he hadn't lost his mind and was still answerable for his actions. It was a dangerous comedy, part of the silly secrecy we had in those days.' The whole procedure was self-defeating. Obviously the envelope had to be placed somewhere within reach in the cabin, and could not be hidden anywhere too hard to find, just in case the need for it was genuine and urgent. An unstable cosmonaut could have opened it at any time without permission and taken control of his ship. Mark Gallai, the Soviet Union's Chief Test Pilot, was recruited into the space programme to help train the Vostok cosmonauts. In a recent interview with historian James Harford, he said:

> All the test pilots believed these concerns were stupid. Many pilots had flown in the stratosphere at night, or in heavy cloud conditions . . . We made a great noise about the key panel. It was our feeling that the chance of a pilot going crazy was much smaller than the possibility of a failure in the radio communication . . . Korolev didn't like the keypad either, but he decided to accept it to quiet the physicians . . . Suppose a cosmonaut made a mistake punching the buttons? Who would punish him?[6]

Curiously enough, the American space pilots fought this battle in precise mirror-image. NASA's cautious rocket engineers wanted fully automatic systems at first, but the astronauts insisted on a wide-ranging freedom of control. They took advantage of their high-profile appearances in *Life* magazine and on television to lobby for command over their own flights, or at least to obtain an equal partnership with their mission managers on the ground. Strong-willed, individualistic astronauts spent long hours throwing their weight around at the various aerospace factories, essentially designing many aspects of the emerging spacecraft to their own convenience.

*   *   *

There was no requirement to prepare any keypad codes or TASS envelopes for unmanned test flights of Vostok, nor to worry about rescue arrangements. If they descended over a foreign country, they could be destroyed by remote control with a 10-kilo explosive charge. If the 'destruct' radio command failed, an on-board timer would blow them up anyway, sixty-four hours after landing. This would discourage American rocket experts from looking into matters that did not concern them.

In fact, the early Vostok capsules needed little help destroying themselves. The first prototype, launched on May 15, 1960, spiralled out of control in space and was lost. Two dogs, Chaika and Lisichka, were put aboard another Vostok on July 28, after the capsule had been modified in the hope of improving it. Now came the R-7's chance to disappoint its makers. Shortly after launch the rocket blew itself to pieces, dogs and all. The Vostok cosmonauts were at Baikonur that day on their first familiarization visit, and they witnessed the launch of the vehicle which, supposedly, was designed to carry them safely into space. Gherman Titov wryly recalls, 'We saw how the rocket could fly. More important, we saw how it blows up.'

On August 19 two more dogs, Strelka and Belka, were sent into space. This time, much to Korolev's relief, the R-7 settled into its climb and the mission proceeded smoothly. Both dogs made it safely back to the ground after seventeen orbits. There was much appreciation in the world's press. Nikita Khrushchev was delighted. Privately, Korolev and the space doctors were disturbed by a small incident during the flight. Belka became dizzy with the weightlessness and vomited into the cabin. Did this mean that humans would also become ill up there? Cameras in the ship recorded the dogs' demeanour throughout. Obviously the journey had not entertained them, but they seemed fine once they were back on the ground.

On September 19, 1960, Korolev formally submitted his proposal for a human flight, and the Central Committee of the Communist Party approved his request. Ten senior figures signed the documents: Korolev; his old ally, the mathematics and computer expert Mtislav Keldysh; the doomed, glory-seeking

Chief of Missile Deployment, Marshal Nedelin; the Chief of Defence Marshal Ustinov; a watchful Valentin Glushko ... If the new adventure was successful, that would mean glory all round. If there was any 'unpleasantness', then ten people could pass the buck.[7]

Korolev planned to launch a cosmonaut by the end of 1960, but Vostok still refused to cooperate. On December 1 another hapless pair of dogs was burned to a crisp when their re-entry ball came down at too steep an angle. On December 22 (the rate of launches was frenetic) a fresh duo of dogs survived an emergency ejection in their special pod, when the R-7 booster ran out of thrust halfway up its climb into orbit. The upper stage's engine did not catch alight and Vostok fell back to earth.

The medical experts exposed many dogs to unpleasant laboratory experiments with barely a second's thought, but the rocket engineers had more feeling for their canine cosmonauts. Yuri Mazzhorin remembers a dramatic race-against-the-clock rescue, in which the space community's concern for one of their animals overcame their fear of the 10-kilo explosive devices in the prototype capsules. 'In 1960, approximately in March, we launched a one-hour flight with a dog. All of a sudden we were advised that the flight was aborted and we weren't getting any more data. Straight away we calculated where the capsule would fall. It was approximately in the area of Tunguska, Siberia region, coincidentally near where a big meteorite fell in 1908. Everyone was upset and said it was a pity the dog would be blown up. Suddenly a signal came through from radio aerials attached to the parachute lines. It meant the ship had survived.'

This was good news, except for a couple of minor details. When they realized the orbit was failing, the controllers had sent up a 'destruct' command. Nothing happened. Obviously the ship was still in one piece when it began its uncontrolled re-entry, but there was no signal to confirm that the dog had escaped in its ejection pod. Perhaps it was still trapped in the re-entry ball? And was the explosive back-up timer activated? If so, the dog would land with a terrible bump, only to be blown up after sixty-four hours!

'Ten people immediately boarded an Ilyushin-14 at Baikonur. There was a bad fog, but they took off anyway.' Cooperative KGB officers were despatched to the more colourful establishments of Samara (then Kuibishev) on the Volga, hunting for a couple of off-duty time-bomb experts with a shared love of drink and girls. 'They were taken from a party, in quite a state, and they were given a plane to Siberia, and we were counting the time left. Perhaps the charge would go off before sixty-four hours? Who knows what the timer was doing? It was a big risk.'

The capsule had come down close to the Arctic Circle. This being March, the daylight in that part of the world lasted no more than a few hours. Fortunately the parachute was spotted from the air just before darkness fell. The bomb was defused and the dog was saved.

In part, this drama was created by the difficulty of maintaining proper radio contact with a spacecraft. The Americans at NASA had the advantage of a worldwide network of listening posts to keep in contact with their Mercury space capsules. They made diplomatic arrangements with Australia, Nigeria, India, the Canary Islands and Mexico to site large and powerful radio dishes on their territory. Communications engineers then laid down an extensive grid of relay towers and undersea cables to connect these stations with the flight managers at Cape Canaveral. (The well-known mission control centre in Houston had not yet been constructed.) In all, the 'Mercury Tracking Network' was a diplomatic and technical achievement just as impressive as the spacecraft itself.[8] It formed the basis of an international system that functions to this day. NASA's spacecraft are never out of communication, unless they disappear for a while behind the moon, or another planet.

Soviet Russia was unable to make such tidy arrangements, because their foreign allies did not live in the right places. Once a spacecraft had disappeared over the farthest horizon of home territory, it was out of communication. The solution was to equip a fleet of four 12,000-tonne cargo ships with special radio masts and send them out into the world's oceans. They transmitted spacecraft data back to Russia, where the signals were in turn relayed to Baikonur for Korolev's inspection. Because the cargo ships' radio pulses were so easy for Westerners to intercept, all the telemetry had to be coded for security. Mazzhorin says, 'Our vessels were observed from the air. The planes came very close, and took many pictures. The [foreign observers] never boarded us, though they probably guessed the ships' purpose from their locations and sailing times. If they did board, the crew were instructed to burn all their code books immediately in a special stove. As soon as each space mission was over, the vessels would carry on and deliver their cargoes – grain, palmira seeds or whatever – to earn money.'

On March 25, 1961, one month ahead of Yuri Gagarin, Ivan Ivanovich flew for the first time, dressed in the same type of spacesuit and equipped with the same model of ejection seat and parachute harness. He flew his Vostok well, and took time to send some radio messages back home, although his observations about space were somewhat strange. In fact, he relayed instructions for making soup: *schi* (cabbage soup) and *borscht* with beetroot and sour cream. The exact details of the recipe are now lost, but it seems to have been a deliberate attempt to confuse any Western listening posts monitoring the flight.

Ivan's descent and landing caused great anxiety for witnesses on the ground. Local villagers saw him come down under his own parachute, and they decided that something did not look quite right. The instant Ivan's feet touched the ground he fell over, apparently unconscious. Naturally the villagers ran over to help, but a cordon of troops quickly surrounded the cosmonaut's prostrate body. The soldiers made no effort to help, but simply stood around him as if to let him die. The villagers were appalled.[9]

In recent times, a sort of Russian 'Roswell' legend has attached itself to this incident. An unacknowledged cosmonaut went up before Yuri Gagarin and was killed during the return phase . . . History was not best served when a pro-communist British newspaper, the *Daily Worker*, published a story just two days before Gagarin's flight, written (or, rather, concocted) by its

Moscow correspondent Dennis Ogden. A renowned test pilot had been injured in a car crash, but Ogden decided that the man was a cosmonaut who had come down to earth badly in a spaceship called 'Rossiya'. As recently as 1979, experts at the British Interplanetary Society took some of these rumours seriously:

> Some controversy surrounds the name of the first man in space. Edouard Bobrovsky, a French broadcaster who visited Moscow in April 1961, revealed that according to reliable sources, Sergei Ilyushin, son of the famous Russian plane designer and a dare-devil pilot, used his influence to go into space himself, three or four weeks before Gagarin. After his return to earth the recovery team found him badly shaken. Sergei Ilyushin has been in a coma ever since.[10]

Actually Sergei was the famous aircraft designer, and his son's name was Vladimir; nor does Bobrovsky sound entirely like a citizen of France. It made no difference to all these rumours that Korolev's launch technicians had daubed MAKET (meaning 'dummy') in thick black paint all over Ivan's face and across the back of his suit, before strapping him into Vostok's ball and sending him off; nor that his soup recipes, beamed back from space, were so obviously the product of a tape-recorder, rather than a live human being. Ivan's choice of subject matter was the cause of heated debate before his flight, as Oleg Ivanovsky recalls. 'We needed to check the radio's ability to convey human speech from space, so we decided to put a tape together. Then the security officials said, "No, because if the Western listeners hear a human voice, they'll think we are secretly flying a real cosmonaut on a spying mission." Remember, this was only a few months [eleven months] after the Gary Powers business. So we thought we'd record a song instead, but the security people said, "What, have you gone mad? The West will think the cosmonaut has lost his mind, and instead of carrying out his mission he's singing songs!" Then it was decided to record a choir, because nobody would ever think we'd launched an entire choir into space, and in the end that's what we did, along with the recipes.'

A less realistic dummy had preceded Ivan on March 9. With these two tests successfully completed, Korolev decided that Vostok was finally ready for a real pilot. He had no choice but to take some risks. NASA's Mercury programme was about to send an American into space. They, too, were prepared to fly brave military volunteers atop missiles with a less-than-perfect launch history, just so long as they could beat the Soviets.

Incidentally, Mazzhorin and his guidance experts had access to many of the documents openly published by NASA, but they also received secret Intelligence reports about forthcoming launch preparations at Cape Canaveral, including the engineering delays and unmanned test failures that dogged the early phases of the Mercury project. This helps to explain why so many Soviet space successes pipped their US equivalents to the post by just weeks, or even days on some occasions. 'I remember once I got this three-page document, data about various secret orbits that the American satellites were following, and I said, "What do I need this for? This is just Newton's Laws of Gravitation." But I reckon our spies had to get those numbers from somewhere. Of course the Americans knew what we were doing, but they stayed silent because we stayed silent. Each side was pretending not to know the other's business. It wasn't a very adult game to play, but it led to great technical progress on both sides and a global space industry with benefits for everybody.'

Quite apart from these complex games of international strategy, the simplest and cheapest security measure that Mazzhorin ever had to organize was for the cosmonauts' benefit alone. 'We put a pistol into Vostok's survival kit, just in case our man landed in the African jungle or some such place, and had to protect himself against wild animals. Not against people, of course. He was supposed to ask any people he came across to help him. He wasn't supposed to shoot them.'

5

PRE-FLIGHT

By the end of 1960 six men from the cosmonaut squad of twenty had been selected as potential candidates for the first Vostok flight. The list was based on the cosmonauts' abilities and their training record over the previous year; however, there was a more arbitrary factor at work – height, or rather, the lack of it. Vostok's ejector seat could only accommodate a crewman of modest stature. Gagarin's short frame made him ideal, as did Gherman Titov's. Alexei Leonov was a highly proficient candidate, but he was too tall for Vostok in its current configuration.

On March 7, 1961 Valentina Gagarina delivered a second child, Galya. Three weeks after this happy event, Gagarin had to leave for Baikonur, where he and Titov were scheduled to rehearse their final pre-flight checks. By now, they were the only serious candidates in the running for the first flight, the list of six having been whittled down yet further. Both men were aware that a final selection for the first flight would not be made until the very eve of launch, scheduled for April 12. Competition was fierce, albeit understated. 'Of course I wanted to be chosen,' Titov explains today. 'I wanted to be the first into space. Why shouldn't I? Not just for the sake of being first – simply because we were all interested to see what was out there.'

Titov and Gagarin tried to outdo each other in their cooperation towards each other, knowing that a spirit of professionalism and teamwork would mark them out as suitable choices. A third potential candidate, Grigory Grigoryevich Nelyubov, miscalculated badly, deliberately trying to push himself forward as the only suitable man for the historic first flight. By the end of March, he was no longer in the running.

On arriving at Baikonur, the cosmonauts' first task was to learn how to dress in their spacesuits. The decision to make the suits had only been taken in mid-1960, after a series of difficult discussions. Many designers thought that Vostok's pressure-shell should be enough to protect its pilot, and Korolev was worried about the extra weight penalty imposed by the suit and its separate life-support system. However, he was swayed by the safety arguments. He turned to Gai Severin, Russia's most experienced maker of pilot garments and ejection systems, and said bluntly, 'You can have the weight allocation [in Vostok] but we need the suits in nine months' time.'[1]

Severin based his suits on the high-pressure aircraft garments he had designed in the wake of the Korean War. The pro-communist pilots in Soviet-built MiGs often lost consciousness if they turned their planes too suddenly during combat, while their American enemies managed to stay awake. Severin realized that a tight pressure-suit could help against the g-forces. After he had dressed the pilots more suitably, the Americans became less keen on chasing MiGs round sharp corners. Using a similar design, his spacesuits would help to brace a cosmonaut against the acceleration of the R-7 rocket. The tight fit, especially round his legs, would prevent blood from pooling in his lower torso and starving the supply to his brain. The strong, airtight layers of the space outfit were made from a tough, blue-tinted rubberized compound, while the outer orange material – familiar to Western observers from publicity photographs – was not particularly important for survival. It was just a coverall to smooth out the various bumps and seals, made from a brightly coloured fabric so that a cosmonaut could easily be located if he came down in a snow-covered region. The Soviet Union in April had many snow-covered regions.

Severin was on hand now to teach the cosmonauts how his spacesuit worked, while the Chief of Cosmonaut Training, Nikolai Kamanin, watched carefully. This lesson was also for the benefit of the attending technicians, who had to be able to handle every component with flawless efficiency, so that nothing would be forgotten on launch day. Two spare outfits were allocated so that the 'real' suits would remain unblemished until they were needed. Then, fully suited up, the cosmonauts took turns clambering through the hatchway of a duplicate Vostok ball, while the pad crews practised strapping them down. Kamanin supervised mind-numbingly repetitive run-throughs of the emergency ejection routine: setting all the right control switches in the cabin, ensuring the suit and helmet were sealed, and above all preparing the body, tensing the muscles, for the violent shock to come. In a real emergency Gagarin and Titov would have to be able to do all these things without a moment's thought.

On April 3 the two rival cosmonauts dressed up in the reserve spacesuits for one last time so that they could be filmed climbing into Vostok. They took it in turns to make a moving farewell speech at the foot of the launch gantry. No clear details of the R-7's appearance were revealed in these shots, because the rocket was still sitting horizontally in the assembly hangar – and its design details were highly secret. The launch technicians mimed the procedures for sealing the cosmonauts up in the ball, these sequences being staged in another area of the main spacecraft preparation hangar, not at the launch pad itself. In the months to come, several faked scenes would be spliced into brief but genuine shots of the launch preparations, taken under much less favourable conditions by cameraman Vladimir Suvorov. On the big day, gantry staff would not be able to give him such full access as he could fake in the hangar.[2]

On April 7 Titov and Gagarin accompanied Kamanin to the launch pad. They inspected the gantry equipment in detail and rehearsed how to get off the pad if a fire broke out. If a cosmonaut was sealed into the ball and something went wrong before the R-7 rocket had even left the ground, the ejection seat would hurl him away from trouble, but at this low altitude he would never get high enough into the sky to open his parachute to its fullest extent. So the engineers had worked out the 'catapulting distance' of the seat and built a huge array of netting on the ground 1,500 metres away from the pad. The cosmonaut would fall into this, just so long as all the calculations were right. A mannequin had made this trip a few times, but now it was for real.

Kamanin reminded the cosmonauts about the manual option. If they were sitting on the pad awaiting lift-off and the blockhouse computers decided that something was wrong with the rocket, then the cosmonaut's seat would automatically eject. Failing that, Sergei Korolev in the control bunker had a special key to activate the seat by remote control, according to his own judgement. Typically he would not trust himself alone. He ordered that two other level-headed people in the bunker should also be assigned such keys. But what if none of these safety options worked properly in a crisis? Then the cosmonaut would have to fire the seat on his own initiative, just like a pilot consciously deciding to bail out of a stricken MiG.

At this point in the lecture, Titov made a casual but most unfortunate remark, as recounted in Kamanin's diary for April 7. 'Worrying about this is probably a waste of time. The automatic ejection system will work without a hitch.'

Kamanin then turned to his other candidate. 'Yuri, what do you think?'

Gagarin considered carefully before answering. Reading his answer, one can assume that he did not want to embarrass Titov or insult the skills of all those engineers who had built the automatic systems, although Kamanin obviously wanted to hear a different opinion. 'I agree, the automatic systems won't let us down,' Gagarin replied, giving Titov some covering fire and expressing proper confidence in the ship's design. 'But if I know that I can eject for myself in case of failure, then that'll simply increase my overall chances.' Kamanin made no particular response, but he carefully noted down the entire exchange:

I kept a close eye on Gagarin, and he did well today. Calmness, self-confidence and knowledgeability were his main characteristics. I've not noticed a single inappropriate detail in his behaviour.[3]

In fact, Kamanin seemed to be having a hard time deciding which man should be the first to fly. Only the day before he had been leaning towards Titov:

> He does his exercises and training more accurately and doesn't waste his time on idle chatter. As to Gagarin, he voices doubts about the importance of the automatic spare parachute release . . . I had already suggested in one of my earlier talks that the cosmonauts make a training ejection from an aircraft, but Gagarin appeared reluctant to do this.

Kamanin seemed to accuse each of the two prime cosmonauts of similar failings with regard to the parachute escape training. Ultimately his final recommendation may have been influenced by a factor beyond his control: the political requirement to favour a farmboy over a teacher's son. However, his diaries suggest a more subtle reason for his ultimate recommendation:

> Titov is of a stronger character. The only thing that keeps me away from deciding in his favour is the necessity to keep a stronger cosmonaut for a 24-hour flight . . . It's hard to decide which of them should be sent to die, and it's equally hard to decide which of these two decent men should be made famous worldwide.

Kamanin obviously believed that Gagarin was capable of flying the single-orbit mission that had now been decided upon for the first manned space flight. He kept Titov in reserve for a more demanding longer flight in the near future. In the circumstances, Titov could not possibly have been expected to see this reasoning as a compliment on his superior discipline.

Some while before he made his fatal mistake with the R-16, Marshal Nedelin constructed a wooden summerhouse at Baikonur as a pleasant change of scene from the usual drab barracks and drearily functional blockhouses. It had an open framework, more like a gazebo than a proper building; a wooden floor; archways, trellises and columns prettily decorated in blue and white. A cool stream trickled nearby. In the cold of winter it was impossible to make sensible use of the building. The airless summer was also impractical, but in April, when the steppe was in blossom for a few weeks and the air was sweet with the scent of wormwood ... There were times when the summerhouse was perfect for a party.

Today, white-haired 63-year-old Gherman Titov bemoans the old summerhouse's sorry state. 'It's windy here now. There were some elm trees, but they cut them down. They should have been replaced, but no one cares. New Russians aren't interested. For them, flying into space is just a business. At least under Nikita Khrushchev cosmonautics was developing. Under the modern Democrats everything just falls down. What's all this history for? Silly fools, they don't understand that when they die, memories of them will also be destroyed. There won't be a single bump left. Not even a grave.'

History is important to Titov, because it was in this summerhouse on April 9, 1961, just three days before the first manned Vostok flight was scheduled, that they celebrated his removal from greatness with vodka, fresh oranges, apples and other splendid foods laid out on a long table. Vladimir Suvorov, the official cameraman, caught the scene on colour film.

The previous day, Suvorov's camera had recorded a more formal event in another part of Baikonur, a special State Committee headed by Korolev, Keldysh and Kamanin, during which the First Cosmonaut was selected. The six prime candidates were standing before them. At the pivotal moment a proud Yuri Gagarin stepped forward to receive his historic commission. In fact, the whole thing was staged. The Committee had already met the previous day, in secret session, with none of the cosmonauts present. Afterwards Nikolai Kamanin summoned Titov and Gagarin to his office and told them, just like that. Gagarin was to be commander and Titov his back-up, his 'understudy'. No explanations. Nothing. Just the awful fact of it, and Gagarin suppressing his usual grin and promising to perform his duties well. Titov says, 'Some people will tell you I gave him a hug. Nonsense! There was none of that. However, the decision had been taken. I understood that.' Kamanin noted in his diary that 'Titov's disappointment was quite obvious.'

Titov mimed his way through the fake State Committee. There was a moment of purest idiocy, a farce within a farce: halfway through Gagarin's carefully pre-rehearsed 'spontaneous' acceptance speech, Suvorov ran out of film. Korolev tapped his glass with a spoon and called the room to silence, as though he had some crucial announcement to make. 'The cameraman needs to reload, so we'll pause for a moment.' Everybody laughed, then sat fidgeting while Suvurov reloaded. Then the First Cosmonaut repeated his earlier performance word-for-word. Meanwhile, Suvorov was struck by Gagarin's youthfulness. 'He was a small, sturdy man, but how young he looked! Like a boy, with a fascinating smile and very kind eyes.'[4]

Next morning there was the more relaxed celebration in the summerhouse, where Titov kept his emotions firmly under control. 'I was upset, of course, but everything went by the script, as they say.' Now he can only wonder if things might have gone differently that day. For he was absolutely convinced that it was going to be him.

Of course the selection of the First Cosmonaut was helped along at the highest levels. Fyodor Burlatsky, Khrushchev's trusted advisor and speechwriter, knows exactly why Gagarin was favoured over Titov. 'Gagarin and Khrushchev were alike in many ways. They had the same kind of Russian character. Titov was more reserved, his smile wasn't so open, he had less charm. It wasn't just Khrushchev who chose Gagarin. It was fate.'

Khrushchev and Gagarin were both peasant farmers' sons, while Titov was middle-class. If Gagarin could reach the greatest heights, then Khrushchev's rise to power from similarly humble origins was validated. Wasn't that the truth? The real reason why they chose a simple farmboy against a properly educated and serious man? After a stiff jolt of vodka to ease the memory and blunt the sharpness of his pride, Titov can now admit, 'I wanted to be the first one. Why not? Many years have passed, and I would like to say they made the right choice. Not because of the government, but because Yura turned out to be the man that everyone loved. Me, they couldn't love. I'm not lovable. They loved Yura. When I visited his mum and dad in the Smolensk region after he was dead, then I realized it. I'm telling you, they were right to choose Yura.'

Gagarin's old academic tutor Sergei Belotserkovsky suggests that another cosmonaut, Vladimir Komarov, came close to being assigned the first flight, 'but a distant member of his family was subject to official repression at the time'. Belotserkovsky attributes Gagarin's eventual selection to a lucky error. 'I was surprised when I found out that Yura's brother and sister had been captured by the Germans. Normally it's a black spot in a person's biography to have lived in occupied territories. Either the vetting authorities missed that, or they didn't take it into consideration. If you like, it was a mistake, but a very useful one. If we could have made more mistakes like that when selecting people for important positions, our country wouldn't have had so many problems. Leaders with an informal attitude to the rules, like Korolev for instance, usually turn out to have the higher standards of morality.'

At 5.00 in the morning of April 11, the doors of the main assembly shed rolled open and the R-7, with Vostok on its nose, trundled into the pre-dawn chill, supported horizontally on a hydraulic platform mounted on a railcar. Korolev paced along the track just ahead, escorting his rocket-child like an anxious parent. The railcar moved at slower than walking pace, so that the rocket would not suffer any vibration damage. All the way to the launch pad four kilometres distant, Korolev never left its side. As Titov explains, 'The rocket was the Chief Designer's baby, if you like. That's why he walked along with it all the way, like a pedestrian. These transports to the pad are very slow. At a time like that, speed is always associated with problems. Vostok rockets are quite delicate as well as powerful – especially that first one.'

At one o' clock that afternoon, Korolev escorted Gagarin and Titov to the top of the gantry for a final rehearsal of boarding procedures alongside the now-vertical rocket. All of a sudden Korolev became weak with exhaustion, and had to be helped down from the gantry and back to his cottage on the outskirts of the launch complex to get some rest. In time, Gagarin would discover that the Chief Designer's stocky, rugged appearance disguised a very fragile man.[5]

Meanwhile, at an Army barracks on the outskirts of Saratov, General Andrei Stuchenko was awoken in the pre-dawn darkness by a telephone call from someone very senior and very frightening at the Kremlin. 'A man is shortly to fly into space. The cosmonaut will land in your district. You are to organize his safe recovery and reception. You answer for this with your head.'[6] Stuchenko promised he would comply. He grabbed a map of his region, divided it into grids and spent the day deploying his troops as fast as he could, to watch for something amazing – a boy falling out of the sky.

The evening before the flight, Titov and Gagarin settled down in a cottage a few kilometres from the pad. Nikolai Kamanin visited them briefly, and (as his diary records) Gagarin took him aside for a few moments, whispering tensely, 'You know, I'm probably not quite right in the head.'

'Why's that?'

'The flight's tomorrow morning, and I'm not the slightest bit worried. Not the tiniest bit, d'you see? Is that normal?'

'It's excellent, Yura. I'm very glad for you. Good night!'

Of course Korolev came along for a few minutes to settle his cosmonauts for the night. 'I don't know what all this fuss is about,' he teased. 'Five years from now, the unions'll be subsidizing holidays in space.' Everybody laughed. Korolev calmly looked at his watch and said good night. This was the signal for the cosmonauts to bed down.

Vladimir Yazdovsky, the senior Director of Medical Preparations, had spent the day in their bunk room, organizing a little treat for the doctors. He had inserted strain gauges into the mattresses to register whether the cosmonauts tossed and turned in their sleep. Wires trailed from the bunks and through a suspiciously fresh hole in the wall to a clutch of batteries outside the cottage. A data cable ran off for a few hundred metres to another building, where the doctors had set up their knobs and dials. Of course this experiment was supposed to be a secret, but as Yuri and Gherman understood from bitter experience, the doctors were sure to demand entertainment of some kind, whatever the time of day. ('There's nobody looking through the keyhole, but you know you're being watched.') History records that both men slept perfectly well. Common sense tells us otherwise. Gagarin eventually admitted to Korolev that he did not sleep a wink. It was not just the impending flight preying on his mind. He wanted to concentrate on lying still, so that the doctors would declare him well-rested and fit for duty in the morning. No doubt his highly disciplined understudy Titov employed a similar trick, with the result that both men were less refreshed next morning than they would have been if the doctors had left them alone. Before going to bed Gagarin confided to Kamanin that he had always considered Titov's chances to be exactly the same as his own. He knew that the merest hint of upset in their last night's 'sleep' could still make a difference. Months later, he joked to Korolev that the only reason he had gone into space on the morning of April 12 was because Titov turned over in his cot the night before.[7]

American Intelligence experts knew perfectly well that preparations for Vostok's launch were under way. Washington time was eight hours behind Baikonur's. While the cosmonauts were resting on their wired-up mattresses, President Kennedy appeared on an NBC early-evening television programme sponsored by Crest toothpaste. He and his wife Jacqueline talked with reporters Sander Vanocur and Ray Scherer about the difficulties of raising their small children, and about the president's 'hands-on' management style. Kennedy mentioned that political events often appeared more subtle and complex from inside the Oval Office than they did to the outside world. Even as he smiled and joked for the television cameras, he knew that a significant defeat awaited him in just a few hours' time.[8]

*   *   *

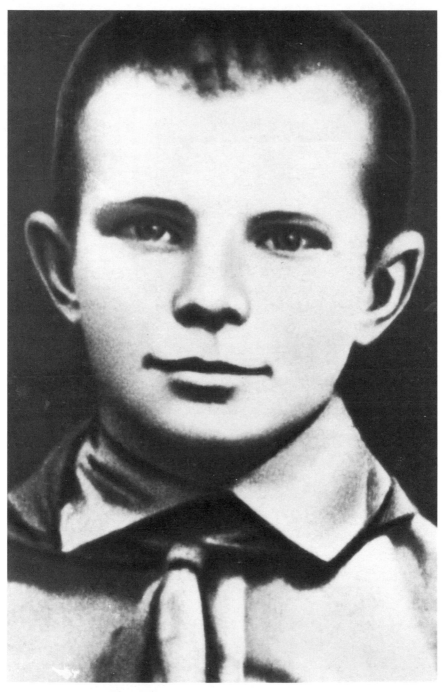

Yuri Gagarin aged six – a child saboteur who watched as Nazi soldiers tried to hang his brother Boris

Lyubertsy
Steelworks: training
as a foundryman,
1950

Gagarin (top right) with other young Soviet Air Force recruits at
Orenburg, 1955

Gagarin during early training in a MiG jet, February 1956

Yuri and Valentina, soon after their marriage in 1957

Yuri and Valentina with their first daughter Yelena, 1960

Parachute training, 1960

Two great
rivals:
Gherman
Titov (left)
and Gagarin

Loyal friend
and fellow
cosmonaut
Alexei Leonov

Gagarin endures the centrifuge, autumn 1960

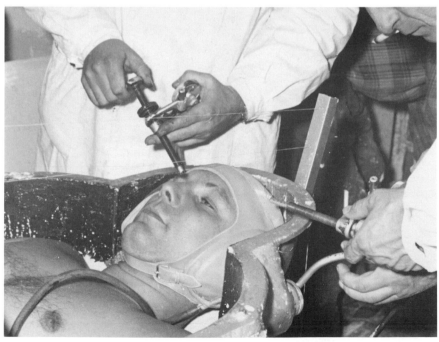

Gagarin being measured for his Vostok capsule seat, March 1961

Suited up prior to flight, 1961

Final preparations, 12 April 1961

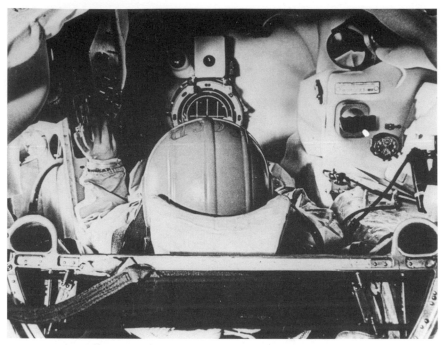

A view through the Vostok capsule's hatch

A Vostok on display showing the troublesome electrical connector

At 5.30 a.m. on April 12, Korolev and Yazdovsky breezed into the cosmonauts' darkened room, all hale and hearty, and turned on the lights. 'What's this, my dears? A lie-in?' Gagarin and Titov went through the motions of awakening from a deep and untroubled slumber.

'How did you sleep?' the doctors asked.

'As you taught us,' Gagarin answered warily.

Korolev went off to check his rocket, but after Gagarin and Titov had washed and shaved, he rejoined them in the day-room for a simple breakfast of concentrated calories and vitamins in the appetizing form of a dark brown paste. Even in Gagarin's heavily censored published account of that day, *The Road to the Stars*, Korolev's exhaustion is evident:

> The Chief Designer came in, and it was the first time I'd ever seen him looking careworn and tired. Clearly he'd had a sleepless night. I wanted to give him a hug, just as if he were my father. He gave us some useful advice about the coming flight, and it seemed to me that talking to us cosmonauts cheered him up a bit.[9]

The doctors arrived from another building across the road to give the cosmonauts a final check-up, bringing with them more of their favourite things: a clutch of sticky, round sensor pads. Titov and Gagarin stood there patiently, half-naked, while the pads were glued into place on their torsos. Star City's director Yevgeny Karpov gave them each a bouquet of flowers to cheer them up. Actually he was passing them on from the old woman, Klavdiya Akimovna, who usually lived in the cottage. He could not let her in just now, which was probably just as well, for she had a touching story about her son, who was a pilot just like Yuri, but who had been killed in the war. No need to mention that, she said. Karpov ushered her away with a few kind words and took the flowers into the cosmonauts' room. He wanted to say something important, something meaningful on this great occasion, but could not think of anything sensible. 'Instead of advice and farewells, all I could do was joke and tell funny stories and other pieces of nonsense, just like everyone else. At the breakfast table we squeezed space-food out of tubes and pretended we thought it was amazingly delicious.'[10]

After breakfast, when the doctors had finished with their pads and glue, the cosmonauts were driven across to the main spacecraft assembly building. The huge construction floor for the Vostok was empty. The rocket and capsule were already out at the pad, but in a closed-off side facility there remained some small but essential items of equipment still to prepare – the spacesuits.

So far, Titov and Gagarin had been treated exactly alike. Now, in the clean white glare of the suiting-up room, a subtle shift of emphasis came into play. Titov was the first to receive his padded undergarment; the first to climb into his pressure-suit; the first into the bright orange outer layer. By the time he was fully dressed in his suit, Gagarin was barely ready. Titov knew not to become excited about this. The technicians had dressed him first, and Yuri second, so that the First Cosmonaut would spend less time between here and the launch pad overheating in his suit. During the drive to the gantry both their suits were ventilated, rather ineffectively, by plug-in fan boxes inside the crew bus. Just in case something happened, Titov had to be ready. A little warmer in his suit than Yuri perhaps, but ready just the same. For now, he knew that getting dressed first meant that he would almost certainly stay second.

Gagarin had his own shock of realization to deal with. In his official account of the flight, *The Road to the Stars*, he wrote, 'The people helping me into my spacesuit held out pieces of paper. One even held out his workpass, asking for an autograph. I couldn't refuse, and signed several times.' Yevgeny Karpov watched Gagarin right up to the last moment. He noticed the First Cosmonaut's anxiety about all these autographs. 'For the first time since his arrival at Baikonur he was at a loss, unable to give his usual instant replies to people. He asked, "Is this really necessary?" I said, "You'd better get used to it, Yura. After your flight you'll be signing a million of these things."' With many months of technical and physical training behind him, Gagarin had not had much time to consider what might happen when he came back to earth after today's mission. Now, almost too late, he caught a glimpse of the enormous social burden that would be placed on his shoulders.

Dressed in the twentieth century's most distinctive suits of armour, Gagarin and Titov took their seats in the bus: a matching pair of cosmonauts, the same sort of age, at the same peak of physical fitness, with the same hard slog of medical endurance and procedural training behind them. Their spacesuits and helmets were identical. It could be either one of them going up today, but by some ridiculous anomaly of fate it was going to be Gagarin. When the bus drew up alongside the base of the launch gantry, Titov wished him good luck, and meant it. Cameraman Vladimir Suvorov recalled the scene in his diary:

> According to our old Russian tradition, on these occasions one should kiss the person going away three times on alternate cheeks. It is completely impossible to do this while wearing bulky space suits with helmets attached, so they simply clanged against each other with their helmets, and it looked very funny . . . Then Gagarin got off the bus and paced awkwardly towards the Chief Designer. Obviously it wasn't easy for him to walk in his clumsy suit.[11]

Titov remained in his seat in the bus, staring listlessly through the windows at the reinforced concrete control bunker – at the 'hedgehogs'. That's what everybody called them: the array of jagged spars sticking out of the roof at crazy angles. The theory was that if a misfiring rocket fell on top of the bunker and exploded, the hedgehogs would break it up before it could actually smash into the roof. The worst of the blast would be deflected, and the people in the bunker might live to launch another day. At least the hedgehogs made more sense than this business of sitting in the bus as Gagarin's back-up. Titov recalls his thoughts that day with painful clarity. 'We'd trained together a long time. We were both fighter pilots, so we understood each other. He was commanding the flight, and I was his back-up, just in case. But we both knew "just in case" wasn't going to happen. What could happen at this late stage? Was he going to catch flu between the bus and the launch gantry? Break his leg? It was all nonsense. We shouldn't have gone out to the launch pad together. Only one of us should have gone.'

Even so, Titov admits that one simple, tantalizing thought went round and round in his head. 'Probably nothing will happen, but what if? No, nothing can happen now, but what if . . .?'

Director of Medical Preparations Vladimir Yazdovsky remembers Titov's palpable tension in the bus. 'Of course he was hoping that when Gagarin went up to the capsule, a small tear would appear in his spacesuit or something, and immediately the Number Two would be in command of the flight, but Gagarin went into the launch-tower lift-cage very carefully, ascended to the capsule and sat down in the cabin, and when he reported to me that he was safely strapped into place, I gave the order to Titov to remove his spacesuit. He answered abruptly, in a disturbed way, but after that he quietened down. He showed no further emotions.'

Korolev, Academician Keldysh and several other dignitaries were at the gantry's base to greet Gagarin and wish him a good flight. 'Well, it's time to go. I've already been inside the ball, to see how it feels,' said Korolev. He took from his pocket a tiny hexagon of metal, a duplicate of a commemorative plaque sent to the moon on a simple automated 'Lunik' probe in 1959. Its deliberate crash-landing had scattered a dozen of the little plaques in all directions. 'Perhaps one day you'll be able to pick up an original, Yuri Alexeyevich.'[12]

In his diary Nikolai Kamanin observed drily:

> When Gagarin left the bus, everybody let themselves release their emotions and started to hug and kiss one another. Instead of wishing him a nice journey, some of them were shedding tears and saying goodbye as if for ever. We had to apply force to pull the cosmonaut out of their embraces.

Then Korolev strode towards the bunker and disappeared under the hedgehogs. Gagarin went on up and Titov stayed down.

After he had ascended to the top of the pad in the lift, the technicians supported Gagarin's shoulders as he raised his legs over the rim of Vostok's hatch and wriggled himself into the ejection seat. Once he had settled, Oleg Ivanovsky and Chief Test Pilot Mark Gallai leaned into the cabin as far as they could and hauled at the loose ends of his straps to tighten him against the seat. Then they plugged his suit hoses into Vostok's life-support system. Gagarin was now an integral component of his ship – or, rather, an integral part of his ejection system. The couch's cylindrical lower section incorporated a pair of solid-charge rocket nozzles, but it also contained a small separate oxygen supply in case the ball sprang a leak in orbit, or Gagarin had to bale out somewhere between the earth and true space, maybe at ten or fifteen kilometres' altitude, where the air is still present but much too cold and thin to breathe.

Down in the blockhouse, Korolev and his technicians saw the life-support monitors flashing their positive signals as the hoses locked into place. Gagarin's air supply was working and his suit showed no signs of leakage. At the foot of the launch gantry, fretting in the bus, Titov received his last orders for the day, standing him down from the mission once and for all.

Fifty metres above the bus and its disappointed cargo, Ivanovsky rapped on Gagarin's helmet with his fist for a final goodbye, but one last detail still troubled him: the keypad codes. 'It didn't feel right to send Yura into space without any real control over his own craft,' he recalls. 'No matter what the psychologists said, he was still a properly trained military pilot.' Surely the whole point of all Gagarin's training was to get him out of lethal emergencies in dangerous craft travelling at colossal speeds? Ivanovsky remembers feeling resentful on Gagarin's behalf. 'The doctors could not judge if his sanity might crack under pressure, because they were not familiar with any kind of flying.' If something went wrong with Vostok's automatic-guidance systems, then surely Gagarin was entitled to flick his own switches and solve the problem his own way, just as he would be expected to pull a spiralling MiG out of trouble without asking permission from a committee of doctors? Vostok was a strange apparatus, but still a flying machine for all that. Just like a plane, it might blow up on take-off, in flight or on landing – Ivanovsky uses the word 'unpleasantness' to cover all these hazards. 'There was always the possibility of unpleasantness to do with flying machines of all kinds,' he says. The only new twist was that Vostok might do none of these things and just quietly carry on in orbit, with the retro-rockets refusing to fire and Gagarin slowly suffocating, with no hope of rescue, no possibility of getting out of the cabin and parachuting gently to earth, his Vostok an eternal tomb . . . Ivanovsky sums up the fundamental risk of a cosmonaut's life: 'His work, his special expertise, may require his death.'

Ivanovsky worried about all these possibilities, although he acknowledges that nobody in the space community ever spoke openly about such things, least of all the cosmonauts themselves. Of his decision on the launch gantry that day, his small rebellion of conscience, he says, 'How should I know why I did it? I must have been undisciplined for a moment.' He leaned through the hatch one last time, signalling for Gagarin to open his faceplate, so that they could talk without using the radio link. Conversations on the wire were not private, and this one certainly had to be. Ivanovsky was about to reveal the Big Secret – the three numbers from the six-digit keypad that Gagarin needed to punch in before he could unlock the spacecraft's manual controls.

'I said, "Yura, the numbers are three, two, five." and he smiled. "Kamanin's already told me." he said.'

Even the hard-hearted Stalinist had been overcome with a dose of humanity at the last moment. As it turned out, so had Gallai and Korolev, although their contempt for the keypad was never in much doubt. Anyway, no more Big Secret. It must have been comforting to know that three other people, including the Chief Designer himself, had broken the rules. In theory Ivanovsky was betraying an official State secret and could have been sent to a prison camp for his crime.

Ivanovsky felt a little happier as he squeezed Gagarin's gloved hand one last time. He and Gallai prepared to seal the capsule, assisted by military Chief of Rocket Troops Vladimir Shapovalov and two junior pad-staffers. First they checked the electrical contacts on the hatchway's rim to make sure that they registered a clear and unambiguous signal. Once the hatch itself was locked down, the contacts would confirm that everything was airtight. They would also prime a series of miniature explosive charges set into the attachment ring, which could blow the hatch at a millisecond's notice, just in case Gagarin needed to eject during a launch failure. The contacts seemed to deliver the right signal, so they manhandled the hatch into position and began to secure the first of thirty screw-down bolts along its circumference. They tightened the bolts in opposing pairs in order to mate the seals evenly.

The instant they had secured the final bolt, the gantry telephone rang. 'We thought this would be Korolev from the blockhouse, ordering us to climb down from the launch platform,' Ivanovsky remembers.

Not quite. It was Korolev, but he sounded far from happy. 'Why aren't you reporting what's going on up there?' he demanded. 'Have you sealed the hatch properly?'

Ivanovsky assured him that they had, just seconds ago.

'We don't have KP-3,' Korolev barked. (KP-3 was the required electrical signal from the contacts on the attachment ring.) 'Can you remove and reseal the hatch?'

Ivanovsky warned Korolev that re-securing the hatch could delay subsequent launch preparations by at least thirty minutes. 'There is no KP-3,' Korolev insisted with his habitual logic. So the hatch had to come off. Ivanovsky thought for a terrible moment how Gagarin might feel when he saw the dawn's early light invading his cabin from a suddenly wide circular hole above his head. 'I said to Korolev, "Can I just tell Yuri? He'll be distressed, and he'll think the hatch is coming off because the flight is cancelled, and we're going to pull him out of the capsule." Korolev said, "Don't worry. Get on with your work in peace. We'll tell Yura."' But Ivanovsky remained agitated. 'In peace? In peace! You can imagine the state we were in. We dedicated our six hands, three pairs, to these thirty little screws, and we had to undo them all with a special key. The hatch panel weighed about a hundred kilos, and it was a metre wide, a massive piece. It wasn't a shameful incident at all, but it was certainly embarrassing.'

Embarrassing and exhausting. Ivanovsky and his colleagues had worked non-stop on Vostok's pre-launch checks, ever since the rocket had reached the pad on the morning of April 11. They had checked the life-support system, the propulsion systems, the navigation gyros: the awesome combinations of electrical energy and explosive chemicals that could miscombine in some small way at any moment and blow Gagarin into pieces (and possibly topple the rocket on its pad, spilling death and destruction across half of Baikonur). They had checked and re-checked, and now this failure of a couple of simple switches on the hatch threatened to ruin everything in the very last minutes before launch. As they pulled off the hatch, Ivanovsky hardly dared look into the cabin. What might Yuri be thinking? 'In fact it was impossible to see his face at that moment. You could only see the top of his white space helmet. Sewn into the fabric of the left sleeve of the spacesuit there was a little mirror, which enabled the cosmonaut to look up at the hatch area, or other areas [of the upper cabin] normally blocked from view [by the rim of his bulky helmet]. He tilted his sleeve so that I could catch a glimpse of his face in the mirror. He was smiling and everything was fine.' Gagarin was whistling quietly to himself as Ivanovsky and his colleagues replaced the hatch.

Those thirty bolts again. Tightening the opposite pairs, and Korolev's voice on the phone, more forgiving now: 'KP-3 is in order.' Ivanovsky does not want to apportion blame, but as far as he can remember, the hatch had looked perfectly good to him right from the start, and he decided that someone in the blockhouse must have made an error reading the data. He would not have minded, except that the KP-3 business gave him a very nasty moment. Now it was the 40-minute mark, and the people in the blockhouse had sent the signal exactly on time to make a partial retraction of the gantries and walkways around the rocket. The platform that Ivanovsky and his four companions were standing on started to move away from Vostok. Any second now it would rotate down to forty-five degrees and they would fall off. There was an awkward moment when they had to use the gantry telephone to request a brief delay in the retraction.

They quickly patted Vostok's ball for luck, then climbed down from the gantry as fast as they could. Their feet were barely on the ground when the hydraulic motors started up again to pull the platforms away.

Ivanovsky headed for the nearest control bunker, while camera-man Vladimir Suvorov opted to stay out in the open, anxious not to miss the most important photographic opportunity of his life. He and his assistants prepared various cameras – manual and automatic – around the pad, only to find themselves being manhandled by soldiers with strict orders to clear the entire area prior to launch. The officer in charge was furious that Suvorov could be so stupid as to stay outside. 'No film crews are allowed. I'm in charge here, and I'm ordering you into the shelters now!' The cameraman's righteous rage turned out to be the stronger. 'We're on official assignment! Here, I'll write you a note,' he sneered. 'I am staying outside by my own wishes. In case of my death you are not responsible, all right?'

'Okay, okay. No hard feelings.' The guards retreated and Suvorov got his historic shots.[13]

The crew bus pulled back from the pad, and Titov was escorted to an observation bunker so that he could strip off his suit. Gai Severin's technicians came at him again like predators, stripping off his gloves, his air hoses and restraint harness, taking him to pieces. He was all flapping arms and crumpled legs, a tangle of stiff fabric, with the neckpiece halfway over his head, when the technicians suddenly rushed away towards the bunker's exit. The launch had started, and they wanted to go outside to see it. 'They forgot about me,' Titov recalls mournfully. 'I was all alone.' He waddled to the exit behind everybody else, clumped up the stairs and emerged onto the observation platform on the bunker's roof.

To this day Titov vividly remembers everything he saw and felt in the next few moments. 'I could hear the high-pitched wine of the fuel pumps pushing fuel into the combustion chambers, like a very loud whistle. When the engines fire up, there's a whole spectrum of sound frequencies, from high-pitched screams to low rumblings.'

He knew the R-7 was hanging over the gulch of the flame trench, suspended on the gantry's four slender support arms. The engines were going through their paces for a few seconds, adjusting to their unimaginable stresses in the last moments before the deliberate inferno of main thrust. 'I saw the base of the rocket belching fire when the engines were igniting, and there were stones and pebbles [from the surrounding scrubland] flying through the air because of the blast.' He watched the claw-like hold-down clamps around the rocket falling away; heard later, much later, the noise too loud to be heard, as the belated soundwave arrived. 'It hammers your ribcage, shaking the breath out of you. You can feel the solid concrete bunker shaking with the noise . . . The light of the rocket's exhaust is very bright . . . I saw the rocket rise and sway from side to side slightly, and I knew from this that the secondary steering nozzles were doing their job properly . . . It's no good describing a rocket launch. Every one is different, and I've seen many. To describe a rocket launch in words is a hopeless task. You have to see it. Every time, it's as if for the first time.'

Titov watched the brilliant fire climb high and dwindle to a spark, a fading impression on the retina, until all that was left was a pungent smoke trail and a silence suddenly much more deafening than the original blast, and all the other people on the platform with their backs turned against him.

Then, back down in the bunker, Gagarin coming through on the radio link, reporting from space. 'It was strange to hear Yuri's voice . . . We were sitting together here just half an hour ago, and now he was up there somewhere. It was hard to understand. Time somehow lost its dimensions for me. That's how I felt.'

Titov on the ground, forgotten. Gagarin up there: the first man in space, his name surviving for as long as human memory survives. And Titov a white-haired businessman and Duma politician in Moscow today, turning up sometimes at Baikonur to watch the paintwork crumble on an old summerhouse, where they took his day away from him.

And if any cosmonaut was more disappointed than Titov, it must have been Grigory Nelyubov, who came within a whisker of making the flight himself. Almost exactly the same age as Gagarin, he had flown advanced MiG-19 fighters as a Navy pilot on assignment to the Black Sea Fleet, before joining the cosmonaut squad in the first group of twenty. A brilliant and intelligent man, his principal fault was his need to be the centre of attention at all times. Although heavily favoured in some quarters to make the first flight, Nelyubov's eventual assignment to third place, behind Titov, came as a great disappointment to him.

Nelyubov's space career did not last – in fact he never went into orbit. On May 4, 1963, Nikolai Kamanin dismissed him from the cosmonaut squad after his drunken skirmish with a military patrol on a railway platform. The patrol arrested him for disorderly behaviour, and (according to Oleg Ivanovsky) he shouted, 'You can't do this. I'm an important cosmonaut!' The military officers agreed to release Nelyubov if he apologized for his rudeness, but he refused. Two other cosmonauts, Anikeyev and Filateyev, were merely bystanders in this drama, but Kamanin sacked them too.

Nelyubov went back to flying MiGs at a remote air station, where he tried to convince his fellow pilots in the squadron that he had once been a cosmonaut and had even served as back-up to the great Yuri Gagarin himself, but nobody believed him. On February 18, 1966, profoundly depressed, he threw himself under a train.

Nelyubov's likeness was airbrushed out of most of the photographs of cosmonaut groups associated with the Vostok programme. In 1973 keen-eyed Western historians discovered a snapshot that had escaped the airbrushing. Star City's political information officers had accidentally released the photo without realizing its significance – and the ungenerous lies that it exposed.[14]

6

108 MINUTES

An hour before the launch, Korolev came on the link. 'Yuri Alexeyevich, how are you hearing me? I need to tell you something.'

'Receiving you loud and clear.'

'I just want to remind you that after the one-minute readiness is announced, there'll be about six minutes before you actually take off, so don't worry about it.'

'I read you. I'm absolutely not worried.'

'There'll be six minutes for all sorts of things, you know.' He meant that a minor instrument problem had created a six-minute delay in the launch sequence.

Then cosmonaut Popovich came on the line. 'Hey, can you guess who's this talking to you?'

'Sure, it's "Lily of the Valley!"'

'Yuri, are you getting bored in there?'

'If there was some music, I could stand it a little better.'[1]

Concerned for every last detail of the flight, Korolev took care of this personally, ordering his technicians to find some tapes or records and set something up straight away.

'Haven't they given you some music yet?' he asked a few minutes later.

'Nothing so far.'

'Damned musicians. They dither about and the whole thing is sooner said than done.'

'Oh, now they've done it. They've put on a love song.'

'Good choice, I'd say.'

8.41. Gagarin felt the shudder of distant valves slamming shut; the rocket swaying as the fuel lines were pulled away. 'Yuri, we're going down to the control bunker now. There'll be a five-minute pause and then I'll talk to you again.'

8.51. The music stopped. Korolev's deep, stern voice on the link, all seriousness now. 'Yuri, the fifteen-minute mark.' This was the signal for Gagarin to seal his gloves and swing down the transparent visor on his helmet. In these last minutes before lift-off there was no NASA-style 5–4–3–2–1 'countdown' on the public-address system (and no public-address system). The rocket would be fired at the appointed instant: 9.06 a.m. Moscow Time. Vostok's guidance expert Yuri Mazzhorin says, 'The Americans only counted down to add drama for their television.' Gagarin's final seconds on the ground were almost anti-climactic.

'Launch key to "go" position.'

'Air purging.'

'Idle run.'

'Ignition.'

All kinds of vibrations now, high whinings and low rumbles. At some point Gagarin knew he must have lifted off, but the exact moment was elusive, identified with precision only by the electrical relays of the gantry's hold-down arms as they moved aside, the four sturdy clamps disconnecting from the rocket's flanks within a single hundredth of a second of each other. Gagarin lay rigid in his seat and tensed his muscles. At any moment something could go wrong with the booster, the hatch above his head might fly away and his ejection charges punch him out into the morning sky like a bullet. This 'life-saving' jolt might kill him – crunch his spine; snap his neck like a chicken's; the hatchway's rim might snag his knees and tear them right off. He had to be prepared.

The g-load climbing. No emergency ejection yet . . . He didn't remember it later, but they told him he shouted out, '*Poyekhali!*' – 'Let's go!' His officially sanctioned account of the lift-off in *The Road to the Stars* clearly shows his fascination with the launch:

I heard a whistle and an ever-growing din, and felt how the gigantic rocket trembled all over, and slowly, very slowly, began to tear itself off the launching pad. The noise was no louder than one would expect to hear in a jet plane, but it had a great range of musical tones and timbres that no composer could hope to score, and no musical instrument or human voice could ever reproduce.[2]

'T-plus seventy.'

'I read you, seventy. I feel excellent. Continuing the flight. G-load increasing. All is well.'

'T-plus one hundred. How do you feel?'

'I feel fine. How about you?'

Two minutes into the flight Gagarin was finding it a little hard to speak into his radio microphone. The g-forces were pulling at his face muscles, 'but really this was not so difficult. The stress was hardly more severe than in a MiG doing a tight turn,' he recalled afterwards. There was a strange moment when all the weight lifted and he was thrown violently forwards against his straps. A shudder told him that the R-7's four side-slung boosters were falling away. The 'Little Seven' paused in its acceleration, as if taking a big breath before the final spurt. Then the central core picked up the pace and the sensation of great weight returned.

At three minutes, the nose fairing fired its pyrotechnic charges, pulling away to expose the ball. Gagarin caught a glimpse of dark blue, high-altitude sky through his portholes. Now he became slightly annoyed at the brightness of the television lamp, which made him squint if his head was tilted a certain way – when he was trying to look out of a porthole, for example.

Five minutes up. Another jolt as the exhausted central core was dropped. Millions of roubles-worth of complex machinery was tossed aside without a second thought, like a spent match flicked to the ground. Vostok climbed the rest of the way into orbit atop a stunted upper stage, with just one small rocket engine. Nine minutes after he had left the pad, Gagarin was in orbit. The vibrations ceased, yet there was no particular sensation of silence. Only those who have never travelled into orbit are in the habit of describing 'the eerie silence of outer space'. The ship was noisy with air fans, ventilators, pumps and valves for the life-support system, and yet more fans behind the instrument panels to cool the electrical circuits. Anyway, Gagarin's ears were covered with microphones hissing with their own special static, or with ground control's ceaseless demands for news. 'Weightlessness has begun,' he reported. 'It's not at all unpleasant, and I'm feeling fine.'

Vostok was rotating gently, partly so as not to waste thruster fuel on unnecessary maneouvres, and partly to prevent the sun from heating any surface area of the craft for too long. Through a porthole Gagarin saw a sudden shock of blue, a blue more intense than he had ever seen. The earth passed across one porthole and drifted upwards out of sight, then reappeared in another porthole on the other side of the ball, before drifting downwards out of sight. The sky was intensely black now. Gagarin tried to see the stars, but the television lamp in the cabin was glaring directly into his eyes. Suddenly the sun appeared in one of the portholes, blindingly bright. Then the earth again – the horizon not straight, but curving like a big ball's, with its layer of atmospheric haze so incredibly thin.

Travelling eastwards, ever eastwards, flying at eight kilometres per second, the dials indicated: 28,000 kmph, although Gagarin would not have experienced any sense of speed.

'How are you feeling?'

'The flight continues well. The machine is functioning normally. Reception excellent. Am carrying out observations of the earth. Visibility good. I can see the clouds. I can see everything. It's beautiful!'

As Vostok swept over Siberia, less than twenty minutes after launch, its steeply tilted orbit carried it to the Arctic Circle, then over the north-eastern hemisphere and towards the North Pacific. At Petropavlovsk on the Kamchatka Peninsula, almost on the extreme east of the Soviet subcontinent, a remote radio monitoring station calculated Vostok's speed and altitude from the incoming telemetry. This would be the final opportunity for accurate measurements before the less well-equipped sea-borne stations took over. Alexei Leonov had arrived at Petropavlovsk a day or two before Kamanin and the State Committee at Baikonur had made the final cosmonaut selection for the flight. As Leonov waited for Vostok's signals, including the crude television picture from within its cabin, he had no idea which of his friends would be in there. 'When Yuri flew, there wasn't a central mission-control complex like the one we have today [at Kaliningrad, north-east of Moscow]. Therefore a number of cosmonauts, familiarized with all aspects of the mission, were disbursed among all the major radio listening posts around the Soviet Union, and as part of that operation I came to Petropavlovsk. I had a small television monitor, and when I saw the picture coming back from Vostok I didn't know if it was Yuri or Gherman, but then I saw some body movements that were characteristic of Yuri. As soon as we made contact he heard my voice and said, "Hi, to the blond man!" That was his nickname for me.'

Petropavlovsk received the telemetry from Vostok, and the signals were laboriously keyed into a secure code, then relayed over ground links to Moscow, where Yuri Mazzhorin, Academician Keldysh and a squad of computer operators unscrambled the codes and fed the data into their gigantic machines.

Petropavlovsk's window of opportunity for radio contact was very brief. Less than thirty minutes after launching into Baikonur's early morning, Vostok swept over the Pacific, into the vast shadow of earth's sleeping half. Down below, the Americans were asleep. Their night-shrouded continents, North and South, sped beneath the ship, and Gagarin would have noted them only as geographers' rumours on his little navigation globe. Now, in this darkest realm, he could see the stars. They were sharp and bright, and did not flicker. There were more stars than he had ever seen from the ground, even on the clearest winter nights.

As its tilted orbit carried Vostok into the southern hemisphere, it sped over Cape Horn and across the South Atlantic. The ground controllers instructed Gagarin to make his switch settings for the re-entry procedure. He checked the 'Vzor' to confirm that the systems had aligned the ship correctly, with the retro-rockets pointing against the direction of travel and aimed above the horizon at a certain precise angle. But Gagarin did not have many switches to alter. It was more a question of telling the controllers on earth what the ship around him was doing of its own accord. As far as any published account shows, he never once touched the controls, or punched the three secret numbers into his keypad.

At 10.25 Moscow Time, seventy-nine minutes into the flight, Vostok's retro-rockets began their deceleration burn precisely on time, just as the craft was sweeping over West Africa, firing for exactly forty seconds, then shutting down correctly. This was the last duty required of the equipment and retropack module behind the ball. The four metal wrap-around straps that held the ball in place were snapped apart by small explosive charges. Gagarin felt the ball twist violently as it came away.

The first phase of the de-orbit burn went according to plan. And in a quiet moment, while the automatic systems ran through their paces, Gagarin began to realize the enormity of what he had done.

> I wondered, 'What will people on earth say when they hear about my flight?' . . . I thought about my mother, and how when I was a child she used to kiss me between my shoulder blades before I went to sleep. Did she know where I was now? Had Valya told her about my flight?[3]

No, his family had not been prepared for the news, because of the secrecy that surrounded the whole mission. Gagarin had been allowed to inform Valya, but he misled her with a white lie, saying that he was going up on April 14, so that she would not worry on the real launch day.

Zoya was getting ready for her shift at the main hospital in Gzhatsk when the news exploded. 'It was very difficult for us. We found it out from the radio. Yuri had told mother he was going on a business trip. When mother asked, "How far?" he said, "Very far", so we didn't know where he was going, or when.' In fact, they had not intended to switch on the radio that morning. Zoya's young son (another Yuri) was doing his homework and needed to concentrate. Anna was quietly cooking. As Zoya recalls, 'Suddenly Valentin's wife Maria came hurrying through the door, out of breath. "Yura!" she said. Mamma became very still. "Tell me, what is it – has he crashed?" And Maria said, "No, not yet." Looking back, it was quite comical, although we were very worried. Maria finally explained, "He's in space!" I lost my temper without thinking. "My God, he's got two young daughters, how did he decide to do that? He must be crazy!" I said.'

But Anna remained perfectly calm. She reached for her coat. 'I'll go to see Valya in Moscow. She'll be alone with the children.'

She was calm, but she wasn't thinking straight, either. Of course she could not just walk out of the door and go to Moscow. She had to get herself to the railway station several kilometres away. Perhaps Valentin could organize a lift for her.

Then they turned on the radio.

Anna put on her quilted coat and best headscarf, then left to see about a train ticket. Zoya contacted her hospital and told them she felt too unwell to work. 'A neighbour came into the house to sit with us, and we listened to the radio. The music with the news reports was cheerful, and we felt a little more at ease. Then the music stopped, and the announcer said that the name of Major Yuri Alexeyevich Gagarin was to be included in the Komsomol Central Committee Roll of Honour. "That's what they do for dead people," I thought.'

Much to Zoya's relief, the TASS radio reports resumed their cheerful tone, and the music came on again, a patriotic march. The announcer said that Gagarin had landed safely. 'It seemed like a huge rock had fallen off my shoulders,' Zoya recalls.

Valentin Gagarin, who lived just next door, was strolling down the road towards work, after spending an hour's lunch break at home, when he heard the news somewhat indirectly. Suddenly his little daughter Olya called after him, 'Papa, come back quickly! Mamma's crying!'

There was a flurry of shouting and rushing from one house to another. Yura was in space, and Anna needed someone to drive her to the station so that she could get to Moscow . . .

For Valentin, it was the parked trucks that made it seem real. He was working in the motor pool. When he reported to the office, he found something strange going on – or, rather, not going on. 'All the trucks were lined up in readiness for their rounds, as usual. The doors of the maintenance shed were open, but none of the drivers were sitting in their cabs. Every engine was switched off. The managers waiting at the other end of the routes weren't calling in to complain about their missing loads. But they always called . . . Nobody in the motor pool was moving. Valentin went to the foreman to see about some time off to take his mother to the station. My boss said, "Can't you see? None of the drivers are working today because they're all listening to the radio about your little brother in space." So I asked: could I borrow a truck for an hour? In his excitement he yelled at the top of his voice, "Yes, take any truck you like. Take the nearest and go!" I climbed into a fuel tanker because it was the first vehicle I found with the keys still in the ignition.'

Valentin picked Anna up near the old electrical sub-station. They had fifteen minutes before the Moscow train was due to leave. As they approached the station there was some confusion with a local motorbike policeman, who wanted to know why Valentin was in such a tearing hurry. When it transpired that he had the First Cosmonaut's mother in his truck, things went more smoothly. The train was already pulling away from the platform, but the stationmaster quickly put out a signal to stop it. 'Mother got aboard, and the ticket officer ran into the carriage also, because in her distraction she'd forgotten to collect her change.'

His nerves thoroughly rattled, Valentin now had to go straight away to the district Party office, because there were many phone calls to answer.

For his part, Yuri's father Alexei Gagarin had left the house very early that morning. There was a job on at the collective farm near Klushino, their old home village. He was contentedly going about his business when another farm worker came up to him and started asking strange questions about his son Yuri. 'What's it to you?' Alexei demanded warily.[4]

'Didn't you hear? On the radio they said that Major Gagarin is flying in space.'

'No, my son's only a Senior Lieutenant. Still, good luck to our namesake, eh?'

All the same, it was a strange coincidence. Alexei thought he might just stroll over to the local Soviet (elected council), for a few minutes and see what was what. When he arrived and popped his head round the door, he found the place crowded and noisy. Local Soviet chairman Vasily Biryukov was busy on the telephone, talking to a district Party official in Gzhatsk. The official was saying, 'We must find out where the cosmonaut was born. Is he in your village records?'

'I don't need any records!' Biryukov shouted excitedly, 'I've got his dad in the room with me! Here, I'll give him the line.'[5]

Startled, unprepared, Alexei took the receiver. Now that he understood what was going on, he was too emotional to speak.

The Gzhatsk officials wanted Alexei to come there straight away, but Gzhatsk and Klushino were completely cut off from each other just now. The spring floods had swept away all the roads. Alexei's lame leg and general health made it difficult for him to walk too far, so Biryukov organized transport by horse over the wettest mud. Then they sat him on the back of a tractor and made a short-cut over the fields, leaving out the drowned roads altogether. One way and another his proud companions got Alexei back to Gzhatsk, where he joined Valentin and Boris in the local Party office, answering a seemingly endless stream of phone calls.

Valentin remembers, 'Each of us was given an office and a phone. A secretary of the Regional Soviet asked, 'For God's sake, can you answer questions about biography? We can't cope with the incoming calls. You see, people will be asking questions that only you can answer.' My brother Boris and I sat down by our phones, and they rang endlessly. There were calls from Moscow, Leningrad, Kiev, Vladivostok and many other towns I'd never even heard of. There were calls from abroad, too. The democratic countries – Romania, Poland, Hungary, all of these.' The switchboard operators had to ration the calls to two or three minutes each. 'At two o'clock the television people came. The district Soviet building was buzzing all day like a disturbed beehive.'

Until this moment, nobody in the world had heard of Yuri Alexeyevich Gagarin, farmboy pilot from the Smolensk region.

As Zoya recalls, 'Of course there was no rest for us. The journalists came from everywhere.' They certainly did. A crowd of journalists with cameras and tape-recorders arrived seemingly out of nowhere. They tried to reach Gzhatsk in Volga cars, Chaikas and Zils, but the rough roads around the town were good only for tractors because of the recent spring floods. The Moscow correspondents had to stop several hundred metres away from the Gagarins' house, then trudge in their city shoes through the mud. Some well-equipped camera crews arrived by helicopter, avoiding the roads altogether. That morning Anna had thrown open the windows of the house to let in the fresh spring air. This was a mistake. A few of the journalists knocked politely on the door and asked to be invited inside, but others clambered in through the windows. Suddenly the house was full of journalists rifling through the Gagarins' every possession, touching, probing, getting things out of place. They were very efficient, almost as intrusive as the KGB, and barely more polite. They asked if any private family photographs might be available and promised to bring them back safely, but they never returned a single one. 'After that, we had no peace,' Zoya says. 'We had telephone calls from everywhere. They all wanted to find out who Yuri was, and where he came from. They didn't know!'

Of course a private telephone was not available to the Gagarins. The only available handsets were in the nearby Soviet community hall, where Alexei and Valentin were busy fielding calls. One of these, later in the day, was from Yuri himself to say that he was all right. Anna had a chance to speak with him that evening, after she'd arrived in Moscow, 'but of course we couldn't quite believe that everything was all right until we could actually see him,' Zoya says. 'You know, we Russians have a saying. You have to touch it to believe it.'

Most published accounts state that Gagarin's descent to earth went smoothly, without serious incident. Gagarin himself was always careful to support this version of events. His official account of the flight, *The Road to the Stars*, contains only a hint of trouble, so fleeting that it was entirely overlooked by Western experts:

> The braking rockets turned on automatically . . . I ceased being weightless, and the growing g-loads pressed me into my seat. These grew and grew, and were heavier than at take-off. The craft began to revolve and I told ground control about it. The turning I had worried about soon stopped and the descent went on normally.[6]

The turning I had worried about . . . The only opportunity Gagarin had to tell the truth, formally at least, was when he testified to a special State Committee, headed as always by Korolev, Kamanin and Keldysh. This meeting was a private opportunity for the cosmonaut to report candidly on Vostok's overall engineering performance during the flight. It was not considered appropriate to release any sensitive technical details to outsiders. Certainly there was no need for Gagarin to tell the world that he could have been killed.

Just before re-entry the ball's main linkages with the rear equipment module separated correctly, but the umbilical cable, with its dense bundle of electrical wires that transferred power and data to the ball, did not come away cleanly. For several minutes the ball and the rear module remained tied together, like a pair of boots with their laces inadvertently knotted. The whole ensemble tumbled end over end in its headlong rush to earth.

The ball was weighted with a special bias, so that the thicker layer of heat shielding at Gagarin's back would swing round naturally of its own accord to face the super-hot onrush of the earth's atmosphere. With the equipment module corrupting the air flow and distorting the proper mass distribution, this alignment was no longer possible. 'The craft began to rotate rapidly. I was like an entire *corps de ballet*,' Gagarin reported to the secret State Committee. 'I waited for the separation but there wasn't any. When the braking rocket shut down, all the indicator lights on the console went out. Then they lit up again. There was no separation whatever. I decided that something was wrong. The craft's rotation was beginning to slow, but it was about all three axes, ninety degrees to the right, to the left . . . I felt the oscillations of the craft and the burning of the coating. I don't know where the sound of crackling was coming from. Either the structure was cracking, or the thermal cladding was expanding as it heated, but it was audibly crackling. I felt the temperature was getting high.'[7]

The heat of re-entry created an ionization layer around the ball, and no voice radio messages could get through. Korolev and his ground controllers probably did not become fully aware of Gagarin's problem until after he had landed.

Atmospheric heating eventually burned through the cable and separated the rogue equipment module, but the effect was to sling the ball away at a tangent with an additional sickening spin. At one point the rotation was so severe that Gagarin began to lose consciousness. 'The indicators on the instrument panels became fuzzy, and everything seemed to go grey.'

Perhaps the State Committee's discussions of this problem did not come soon enough for the engineers to make suitable adaptations, ahead of Gherman Titov's mission? At any event, he survived a similar difficulty when he flew on August 6, 1961. Gagarin's post-flight description of the separation failure was perfectly calm, quite relaxed, but Titov says that if his own experience was anything to go by, he must have wondered: 'Which is stronger, the capsule or the other module? Which will break first? You switch on all the recorders and transmitters, to try and report in case you don't make it. You see the little earth globe rotating, and the clocks still running, which means information is still coming from the equipment module through the cables. The capsule rotates very fast. Then there's a huge shaking. Both compartments are hitting each other. Is it scary? That's an interesting question. I could have been scorched, but so what? Similar things have happened.'

At last Gagarin heard denser air whistling past the ball and his whirlwind rotation became less severe. Outside the charred porthole he saw pale blue sky. He was shaken, but any minute now he knew that further stresses awaited him. At seven kilometres' altitude the hatch above his head blew away. The noise was terrible. The cabin seemed suddenly so very open, so exposed. According to Gagarin's published account, he wondered for a crazy moment: 'Was that me? Did I eject just then?'

His account does not quite square with the recollection of Vladimir Yazdovsky, Korolev's Director of Medical Preparations, and a member of the ground-control team at the time. He remembers Gagarin triggering the ejection himself.

The entire procedure was supposed to be automatic. When the pressure sensors registered an atmospheric pressure consistent with an altitude of seven kilometres, Gagarin would come shooting out of the ball, and at four kilometres, the ejection seat's propulsion pack and large parachute canopy would fall away, releasing him so that he could descend more gently under his smaller personal parachute. If his seat did not fire at the right time of its own accord, then he had the option of triggering the ejection himself, but he was not supposed to do this without good reason.[8]

As the ball began to slow down in the denser atmosphere and the heat of the initial re-entry faded away, Gagarin's radio link with ground control was restored. According to Yazdovsky, 'He reported that the g-loads were still very heavy, and they were pulling him in different directions. We said, "Hang on in there." We suggested to him not to eject too soon, but he ejected early, from an undefined height.'

It seems that the ground controllers were unaware of the separation problem that Gagarin had encountered earlier and did not realize why he was complaining about excess spin and g-loading. Perhaps Gagarin did not have time to explain in greater detail; or perhaps he knew that he should not discuss the separation problem over the voice link, in case any Western listening posts were eavesdropping. The dialogue in this very last phase of the mission has never been published, but the historian Philip Clarke believes that the ball might still have been rotating at an uncomfortable rate, long after the equipment module had finally separated. Gagarin's decision to eject early was not necessarily a panic reaction. He may have believed that the spinning of his capsule would interfere with his ejection, and the sooner he attempted it, the better.

In the event, Gagarin's ejection and touchdown went smoothly. As soon as the ejection seat's rocket charges were spent, a large parachute canopy unfurled to slow down his fall. Then the seat fell away, as planned, leaving him to drift more gently to the ground under his own parachute.

Baikonur's morning was Washington's night. At 1.07 Eastern Standard Time, American radar stations recorded the launch of an R-7 rocket, and fifteen minutes later a radio monitoring post in the Aleutian Islands off Alaska detected unmistakable signs of live dialogue with a cosmonaut. White House science advisor Jerome Wiesner called President Kennedy's press secretary Pierre Salinger with the news. Salinger had already prepared a statement for Kennedy to read out. The President had gone to bed a few hours earlier with a sense of foreboding. Wiesner asked him whether he wanted to be woken as soon as the rocket was launched? 'No,' the President answered wearily. 'Give me the news in the morning.'[9]

At 5.30 a.m. Washington time, the Moscow News radio channel announced Gagarin's successful landing and recovery. An alert journalist called NASA's launch centre in Florida to ask if America could catch up. Press officer John 'Shorty' Powers was trying to catch a few hours' rest in his cramped office cot. He and many other NASA staffers were working 16-hour days in the lead-up to astronaut Alan Shepard's first flight in a Mercury capsule. When the phone at his side rang in the pre-dawn silence, he was irritable and unprepared. 'Hey, what is this!' he yelled into the phone. 'We're all asleep down here!' Next morning the headlines read: 'SOVIETS PUT MAN IN SPACE. SPOKESMAN SAYS US ASLEEP.'[10]

On the afternoon of April 12, President Kennedy held a press conference in Washington. Normally a self-confident and eloquent public performer, he seemed distinctly less sure of himself than usual. He was asked, 'Mr President, a member of Congress today said he was tired of seeing the United States coming second to Russia in the space field. What is the prospect that we will catch up?'

'However tired anybody may be – and no one is more tired than I am – it is going to take some time. The news will be worse before it gets better. We are, I hope, going to go into other areas where we can be first, and which will bring perhaps more long-range benefits to mankind. But we are behind.'[11]

7

COMING HOME

Korolev, Mazzhorin and the trajectory mappers at OKB-1 knew precisely the direction that Gagarin's ball would adopt as it plunged through the atmosphere and fell to the ground. What they did not know was exactly how far along that direction it would travel before coming to rest. The trajectory calculations from Mtislav Keldysh's computers were good to an accuracy of just a few kilometres. In the vastness of outer space this was more than acceptable. For a homecoming Vostok ball it could have meant the difference between landing harmlessly in an empty field and crashing through a roof, killing all the people beneath. With great care, Vostok's incoming route was selected to place as few houses as possible in the path of danger. All the descent scenarios favoured large meadowlands, scrublands and fields.

Today, Russian capsules come down onto the vast (and supposedly uninhabited) steppes of Kazakhstan, not far from where they lifted off in the first place. A well-rehearsed procedure for capsule location and crew retrieval has operated for three decades. Back in 1961, Korolev and his mission planners were not quite so ready to dump their very first cosmonaut into the middle of nowhere. Gagarin fell to earth only a short distance from where he had first flown an old Yak-18 at the Saratov AeroClub thirteen years earlier. The exact location of his touchdown was twenty-six kilometres south-west of the town of Engels in the Saratov region, on the outskirts of a village called Smelkovka.

From ground level there was no possibility of observing the ball's hatch flying away, or the sudden jolt of Gagarin's ejection seat. At seven kilometres' altitude, this was all happening too high up to be seen. But tractor driver Yakov Lysenko heard a distinct crack in the skies above his head. Naturally he looked up. The faint echo of the hatchway's explosive bolts took twenty seconds to reach him. By that time Gagarin and his craft had fallen three kilometres closer to the ground, and their parachutes had opened. They were just about visible now to the naked eye. In fact, it is probable that Lysenko heard a different bang closer to the ground, when the ball's parachute hatch was blown off at just four kilometres' altitude to deploy the folded canopy from within. 'You can hear an explosion if it's a plane or something like that, but I saw there was no plane,' says Lysenko. 'There was no engine roar. I was standing and watching, and I saw a ball in the air. Well, not a ball, but something landing with a parachute. A pilot from a plane, I thought.'

Lysenko ran back to Smelkovka village to raise the alarm. He gathered together a reassuring group of friends, and they all tramped across the fields to the spot where he had seen the 'pilot' come down. Gagarin seemed very happy to encounter ordinary folk like them. 'We came to the place, and he was coming towards us. He was very lively and happy, especially after he landed successfully. He was wearing a jump-suit or whatever it's called, and he said, "Boys, let's be acquainted. I am the first space man in the world, Yuri Alexeyevich Gagarin." He shook hands with everyone. I introduced myself, and he said, "Boys, don't leave. All the bosses will be here any minute now. They'll come by car, lots of people, but don't leave. Let's take a picture so we'll remember this." But of course everyone forgot about us. They came from a city or a military garrison. They took him into a car straight away. He told us not to leave, but they drove him away, and we've never seen him since.'

The official reception party arrived with terrifying speed, as if out of nowhere. General Stuchenko (his head still at risk) had been monitoring the skies all morning with close-range radar. Gagarin and his re-entry capsule were located long before they hit the ground, and Stuchenko disbursed his forces accordingly. 'The military came by plane. Some of them were even landing by parachute,' Lysenko recalls. 'It was a complete invasion force.

They didn't allow us to get too close. They're very strange people, you know.'

Lysenko may not be a sophisticated man – he's just a simple tractor driver – but his grasp of the geopolitical significance of what he saw that day makes for a fine summary. 'The Soviet Union announced a spaceship, the first in the world, with Yuri Alexeyevich Gagarin. The whole country was rejoicing. You know, it was a shame to the foreign countries. America is a mighty country, but they didn't quite make it to be first. As they say, "It's important who crosses the bog first." That's how I understand it.'

Lysenko and his friends were not the only people to see the cosmonaut come down, as Gagarin's sanctioned account of his landing makes clear:

> Stepping onto firm ground again, I caught sight of a woman and a little girl standing near a dappled calf and looking curiously at me. I was still in my bright-orange spacesuit, and they were a bit frightened by its strangeness. 'I'm a friend, comrades! A friend!' I shouted, taking off my helmet and feeling a slight shiver of nervousness. 'Can it be that you have come from outer space?' the woman asked. 'As a matter of fact, I have!' I replied.[1]

A *slight shiver of nervousness.* Everyone in the Soviet Union knew about the American spy Gary Powers, who had been shot down over Russian territory the previous May. Maybe this orange-clad pilot was yet another foreign spy parachuting from his stricken aircraft? A number of Western aerospace historians believe that some of the farm workers came at Gagarin with raised pitchforks, only standing down their weapons when they caught sight of the big red letters 'CCCP' emblazoned on the upper front of his white space helmet. Today the Soviet space journalist Yaroslav Golovanov is prepared to admit that 'when they saw Gagarin's orange protective suit, the women became frightened, because there was all this business about Powers only a year before. They said, "Where are you going? Where are you off to?" They thought maybe he was a spy.'

TASS radio announcements of the flight had been broadcast well in time for Gagarin's actual descent. In all likelihood the farm workers greeted him with nervousness at first, but not with outright hostility. It may be that some of them had left their houses early that morning to go to work in the fields and may not have heard the radio bulletins about the space flight . . .

So who exactly had the opportunity to greet the cosmonaut first? Was it Lysenko and his pals, or the woman and child whom Gagarin spoke of? 'Oh, I forgot about that,' says Lysenko. 'Yes, when we went to where he landed, Takhtarova, the local forest warden's wife, was weeding potatoes with her grand-daughter. They had a small piece of cultivated land nearby. When he landed, we were not there. She was scared, and wanted to run away. Then he saw us.'

Later that day a simple signpost was erected at this site, more or less where Gagarin's feet had touched the ground:

DO NOT REMOVE!
12.04.1961
10.55 MOSCOW TIME

Two days later a more permanent stone obelisk was erected bearing a plaque: 'Y.A. Gagarin Landed Here.' No similar marker identified the place where his empty spaceship came down. Gagarin's ejection at high altitude had caused him and his capsule to drift apart by more than two kilometres by the time they hit the ground. Contemporary documentation blurs the whole issue. Recording the capsule's landing site would have meant admitting the forbidden secret of the First Cosmonaut's separate descent under his own parachute. However, the exact site is known – unofficially at least – because a group of children playing in a meadow near the banks of a tributary of the River Volga saw the empty ball come down, alarmingly close to a ditch. It made a dent in the soft ground. Today that hollow looks like a thousand similar indentations in the gentle surrounding grasslands.

Two schoolgirls, Tamara Kuchalayeva and Tatiana Maka-richeva, ran over to see this amazing object. 'We were supposed to be in a lesson at school, but all the boys ran off. They saw a ball flying,' says Tatiana. 'It was huge. It fell down, then bounced and fell down again, settling on its side. There was a large hole [in the ground] where it fell for the first time. The boys ran to it and climbed inside. They picked up many small tubes of cosmonauts' food and brought them back to school, and they told us the ball had landed.' The two fit, handsome women are mildly surprised that their nostalgic hike across the hillocks and furrows to see the landing place now makes for quite a sturdy walk. 'Today we come here and are already tired, but at that time you can imagine how fast we children ran!' says Tamara. 'We'd heard the radio announcement and we all ran with inspiration.'

Proudly the boys handed out the tubes of space food they had found. 'Some of us were lucky and got chocolate,' Tatiana recalls. 'The others got mashed potatoes. I remember tasting some and spitting it out.'

Tamara says, dismissively, 'If you offered it to us today, we wouldn't eat it.'

By now the children (and a good many adults besides) were clambering in and around the ball looking for souvenirs. The military security squad had arrived, though not yet in sufficient numbers. According to Tamara, 'They tried to scare us off. "Go away, go away!" they said. "It could explode!" Their threats didn't have the slightest effect on us.'

Actually there were several opportunities for citizens of the Saratov region to collect souvenirs. Gagarin had cut loose from his parachute the moment he landed, because he was slightly worried that the wind might drag him off his feet. That parachute went missing soon afterwards, while the ball's larger canopy was shredded by souvenir-hunters. The cabin's heavy hatch came down somewhere, as did a detachable radio transceiver and other items of survival gear, and the second hatch covering the ball's parachute compartment.

All these components had some interesting adventures before they were recovered. For instance, there was the matter of the raft. If the ball had come down over the sea, Gagarin might have stayed with it until splashdown, because the impact on water would have been less severe than on the ground; but the ball was not guaranteed to stay afloat for an indefinite time, so he would have clambered with all due haste into an inflatable life-raft. In the event, he came down over Russia precisely according to plan, and the raft stayed packed inside his survival pack. Apparently someone removed it without authorization, and a day or so later he took it to the nearby Volga tributary to do some fishing. A large detachment of KGB officers arrived in the district and requested that all stolen equipment from the Vostok be surrendered, including the raft. They threatened the entire population of Smelkovka with detention if the missing equipment was not returned immediately. Tractor driver Lysenko remembers a certain degree of fuss about 'something being torn, or going missing. Perhaps it's better not to say . . . Some of our boys, the younger ones, found a boat. The special police came and said, "It belongs to the State. We must have it." They visited all the houses and put pressure on various people.'

Yaroslav Golovanov adds his contribution to the story. 'Eventually the KGB recovered their prize, and the unhappy fisherman could only fidget and say, "I'm sorry, but the boat is dilapidated, ripped apart." The KGB officers wanted to correct the fisherman's false ideas. "The boat is fine. Nothing has been torn," they said.'

Apparently the KGB officers did not want to tell their superiors that various historic items from the First Cosmonaut's equipment had been tampered with, before they could collect them all.

Gagarin's social responsibilities began the instant his feet touched the ground. The old woman and the little girl needed reassuring that he was not an enemy spy. He really wanted the farm lads from Smelkovka to be rememembered, because they had been so friendly. Now the military was all over the place, and the officer on the scene, Major Gasiev, came up to him. Gagarin saluted smartly and said, 'Comrade Major! USSR Cosmonaut Senior Lieutenant Gagarin reporting!'

'Listen, you're a Major too. Don't you know? You were promoted during your flight,' said Gasiev with a big grin. They embraced amicably, officers of equal rank, and of course Gasiev had a hundred questions.[2]

Then there was the question of the Altitude Record. The sports official Ivan Borisenko needed Gagarin to sign some documents. In a 1978 account Borisenko described 'dashing up to the descent module, next to which stood a smiling Gagarin'. This seems unlikely because the capsule was at least two kilometres away, and Borisenko must have rendezvoused with the cosmonaut at his separate touchdown site, or else in another field on the outskirts of Smelkovka, where a large helicopter was sitting ready to take Gagarin to the nearby Engels airbase. Some time very soon after his landing, the First Cosmonaut blithely put his signature to a sheaf of Borisenko's off-white lies.[3]

In the helicopter, Gagarin politely and enthusiastically answered all the questions that his military escorts threw at him. What was the earth like? The weightlessness? He was learning that all the questions would be similar, wherever he went. But at one point he went quiet for a moment. According to Golovanov, he said, 'You know, I never got to see the moon through my porthole . . . Never mind, I'll see it next time.' With that, he brightened up and took more questions.

General Stuchenko, his career resting safe for today, met Gagarin on the tarmac at Engels and immediately posed another complex social challenge for the space traveller, as witnessed by Golovanov. 'Yuri Alexeyevich, in the battles to liberate the Gzhatsk district, there was only one commander. You must remember me?'

'No, I don't.' This was not such a good reply. Stuchenko looked absolutely crestfallen, and Gagarin had to think fast. 'I mean, I don't remember your face. But I remember there was a commander. So that was you? How wonderful! You must be my double godfather. Once you rescued me from the Nazis, and now you're meeting me on my return from space!'[4]

This response proved more satisfactory. Then Stuchenko asked, 'How would you like it if we sent a plane to Moscow to fetch your wife? Valentina could come here and then you could fly home together.'

Another awkward problem: how to refuse such a kind offer from a superior officer, and a General, no less. 'Thank you very much for the thought, Comrade General, but I'm afraid it wouldn't do. Valya's nursing our newly born daughter at the moment.'

Stuchenko escorted Gagarin into the airbase's officer's quarters, where he had an opportunity to call his family and report his success to the First Secretary on a secure phone link. Gagarin spoke carefully, knowing full well that his every word would be written down for posterity.[5]

'I'm glad to hear your voice, Gagarin Alexeyevich.'

'Nikita Sergeyevich, I'm glad to report that the first space flight has been successfully completed.'

Khrushchev continued in an official vein for a while, but he could not long resist the ordinary questions. 'Tell me, how did you feel in flight? What's space like?'

'I felt fine. I saw the earth from a great altitude. I could see seas, mountains, big cities, rivers and forests.'

Now, for Khrushchev, the real fun. 'We shall celebrate together with all the Soviet people. Let the world look on and see what our country is capable of, the things our great people and our Soviet science can do.'

Gagarin dutifully echoed the sentiment. 'Now let the other countries try and overtake us.'

'Exactly! Let the capitalist countries try to overtake us!'

According to his senior aide Fyodor Burlatsky, Khrushchev was deeply impressed by Gagarin's cheerfulness and the enthusiastic nature of all his replies. He genuinely looked forward to seeing the young man in Moscow for a splendid and very public celebration in two days' time.

'Thank you, Nikita Sergeyevich,' Gagarin signed off. 'Thank you again for the confidence placed in me, and I assure you I'm willing to carry out any further assignment for our country.'

He was thinking about the moon, perhaps. After all, Korolev had pressed a replica of a lunar spaceship's plaque into his hand just before he took off and said that one day he might pick up the original . . .

The halt at Engels was little more than a rest-stop, a chance to make those necessary phone calls and switch to a more powerful aircraft. Gagarin boarded an Ilyushin-14 for Kuibishev (today called Samara), another large town on the Volga about 350 kilometres north-east of Saratov. Here he would rest for a day or so, before heading for Moscow early on April 14.

An hour after Gagarin's rocket had left the launch pad, Titov, Gallai, Kamanin and a substantial delegation from Baikonur boarded an Antonov-12 plane bound for Kuibishev. Korolev was the most notable absentee from the Antonov's passenger list. He was still monitoring the communications from distant radio ground stations and listening ships, which were tracking the final phase of Vostok's orbit and descent. (He and Ivanovsky flew out later to supervise the ball's recovery from the Saratov region.)

Titov's mood was strange, perhaps rather surly. 'We landed at Kuibishev airbase, which was also a big factory facility for making passenger aircraft. Then Yura arrived by Ilyushin-14. He was brought from the Saratov area. He was surrounded by Generals, and I was just a Senior Lieutenant with very small shoulder-straps, as they say. But I was interested to know: what was the weightlessness like? Yura was walking down the gangway, and I pushed everyone aside. All of them looked at me. "Who's this lunatic Lieutenant?" they said. We other cosmonauts were top-secret [unknown] people, so to say. But I reached Yura. "How was the weightlessness?" I asked. "It's all right," he said. That was our first meeting after his flight.'

The airfield's perimeter fences were buckling under the weight of curious onlookers, who all knew what was afoot. Kuibishev was seething with exultant crowds when Gagarin's car left the airbase and passed along the city's main street escorted by motorcycle police and so forth. 'Someone in the crowd threw a bicycle under the car's wheels because they wanted Yura to stop and say hello. The car swerved to avoid an accident,' says Titov, who was in the car behind. 'I don't recall if the bicycle was damaged, but people wanted very much to see him.'

On the outskirts of Kuibishev, a special *dacha* on the banks of the Volga had been prepared so that Gagarin could take a medical examination and get a day's rest before flying to Moscow early on April 14. Oleg Ivanovsky remembers meeting him there and giving him a huge hug. 'I asked him, "How are you feeling?" and Gagarin replied, "What about you? You should have seen yourself at the launch pad when you opened the hatch. Your face went every colour of the rainbow!" Everyone was rushing up to him, but I didn't lose my head. I gave him that morning's newspaper, and he wrote me a few kind words alongside the front-page photograph of him wearing his helmet. Everybody involved with Vostok was coming up to him and asking: did he have any comments about their particular items of equipment?'

Somehow Gagarin managed to find time for a shower, a semi-quiet stroll by the Volga and a proper meal, all the while being charming and helpful to his endless interlocutors. At a preliminary press conference he gave his impressions of the earth from space: 'The day side of the earth was clearly visible – the coasts of continents, islands, big rivers, large surfaces of water. I saw for the first time with my own eyes the earth's spherical shape. I must say, the view of the horizon was unique and very beautiful.' Then he described a sunset as it appeared from orbit, and the incredible delicacy of the earth's atmosphere seen edge-on. 'You can see the colourful change from the brightness of the earth to the darkness of space as a thin dividing line, like a layer of film surrounding the earth's sphere. It's a subtle blue colour, and the transition is very gradual and lovely. When I emerged from the earth's shadow there was a bright orange strip along the horizon, which passed into blue, and then into a dense black.'[6]

Some correspondents (and even a few of the cosmonauts) found it hard to come to terms with Gagarin's eloquent description of a sunrise and sunset all happening within the space of ninety minutes (Vostok's basic orbital period, not allowing for the boost and deceleration phases). Nor did many people understand what he meant by the 'earth's shadow'. The adventure of space flight – so commonplace now – seemed very magical and strange in 1961.

In the evening, when all but the most intimate colleagues had at last been sent on their way, Gagarin played a quiet game of billiards with Cosmonaut Two: a polite but subdued Gherman Titov. 'I still feel jealous, right up until now,' Titov admits. 'I have a very explosive character. I could easily say rude things, offend someone and walk away, but Yuri Alexeyevich could talk freely to anyone – Pioneers [boy scouts], workers, scientists, farmers. He could speak their language, you see? I was jealous of it.' Yet they were pilots both. They would always have that in common. And mutual respect, if not an absolute love. They played billiards and Gherman listened with genuine interest as Gagarin explained certain events of his flight. With that success now stamped into the pages of history for ever, at least Cosmonaut Two could be sure of his own chance of flying in space during the next few months. Vostok was proven, and Korolev's 'Little Seven' seemed to be working more reliably now, after its somewhat frisky adolescence. However, the *dacha's* billiard room may have been too public a place for Gagarin to go into details about the equipment module's failure to separate properly. That was an unpleasantness that Titov would have to discover for himself. He does not specifically remember Gagarin warning him about it in advance.

One persistent journalist snapped some informal shots of Gagarin under the dim lights of the billiard room before being sent on his way. 'Surely that's enough, now,' Titov said.

It wasn't nearly enough.

Next day Korolev, Kamanin, Keldysh and the other members of the State Committee convened at the *dacha* and took evidence from Gagarin about his flight. Behind closed doors he felt free to describe the retro-pack problem in detail. To this day there is no clear explanation as to why the issue wasn't resolved in time for Titov's flight on August 6. Probably an alteration was made, but it simply did not work. The data cables from the rear equipment module slotted into a large round plate on the ball, through a plug with seventeen pins, each consisting of an array of smaller pins, so that eighty separate electrical connections were made in all. It was no simple matter to eject such a complicated plug. These and similar basic mechanical problems dogged the early years of Russia's space effort.[7]

In America, NASA's engineers also recognized the difficulty of separating re-entry capsules after a flight. Like the Russians, they relied on thick bundles of wires for connecting the capsules to their support modules. The crucial difference is that if their plugs failed they could simply slice right through the umbilical bundles with a guillotine blade powered by a small explosive bolt. If that blade failed, another one further up the cable took over. The flexible wires were so much easier to get rid of than the bulky plugs. Uncharacteristically, the brilliant Korolev did not anticipate this solution, and Vostok's connector caps were over-elaborate, with drop-away clips, mini explosive bolts and other mechanisms that did not quite work on the day.

Early on the morning of April 14 Gagarin left for Moscow. He climbed the gangway of a large Ilyushin-18 airliner capable of long-distance flights. In a few weeks' time he would no longer think of this sturdy machine as an IL-18 any more; with weary humour he would be calling it 'home'.

Much of his time during the flight to Moscow was taken up with journalists' questions. Messages of congratulation were coming in via the radio in the cockpit, and the aircraft's crew took turns coming into the passenger cabin for a few words. As the plane approached its destination after four hours' steady progress, Gagarin took time out to gaze through his window. He saw his old life shooting past him in the sky, and a new and more complex life waiting on the ground:

We were escorted towards our landing in Moscow by a squadron of fighter planes. They were lovely MiGs, just like the ones I used to fly. They came in so close to us that I could clearly make out the pilots' faces. They were smiling broadly, and I smiled back. Then I looked below and gasped. The streets of Moscow were flooded with people. Human rivers seemed to be flowing in from every part of the city, and over them, sail-like red banners waved on their way towards the Kremlin.[8]

The Ilyushin touched down at Vnukovo Airport earlier than expected. Gagarin had to stay aboard for a few minutes until the pre-planned schedule of celebratory events was due to start. He felt happy but nervous.

Down on the ground, Valentin, Boris, Zoya and Alexei had met up earlier with Nikita Khrushchev and his wife Nina in Moscow, where they had already caught up with Anna and Valya. Zoya remembers a great deal of kindness from the First Secretary and his wife. 'He was very simple and down-to-earth with us, and she spent all her time with us. We spent four days in Moscow, and every morning Nina would come round to us and only leave in the afternoon. It was a very liberal situation.'

The first formal event took place at the airport. 'We stayed in Moscow to have some rest, and then on the fourteenth they took us to Vnukovo to meet Yura. We had just arrived, and we saw a plane with an escort of fighters, and they told us it was Yura's. But when it landed he didn't come out for some while, so we started to worry. Nina Khrushcheva said, "Don't worry, the plane arrived a little earlier than planned, but as soon as the time comes, your Yura will come out." And truly, in a few minutes he came out.'

They had laid out a long red carpet. (Nina Khrushcheva told Valentin, 'Usually it's a blue one.') Yuri walked down the gangway and onto the carpet, looking every inch the hero in his brand-new Major's uniform and greatcoat, but Zoya immediately noticed something terrible. 'I saw something dragging on the ground behind him. It was one of his shoelaces.' Gagarin noticed it too, and spent the interminable ceremonial walk along the carpet silently praying that he would not trip over and make a fool of himself on this of all occasions. He told Valentin later that he had felt more nervous on the carpet than during the space flight. But he did not trip. Incidentally, the shoelace can be seen in the many commemorative films of the day's events. The cosmonauts' official cameraman, Vladimir Suvorov, noted in his diary the endless discussions later about whether or not to edit the film and remove the scenes showing the untied shoelace. Eventually, at Gagarin's insistence, the shots were preserved as a sign of his ordinary, lovable humanity. The 'mistake' turned out to have its own special propaganda value.[9]

A smiling, sure-footed Gagarin reached the flower-decked reception platform in one piece, greeted Nikita Khrushchev and other senior Party officials, then hugged his family. Valya gamely awaited her turn in the queue for a hug and a kiss. Alexei and Anna were dressed in their simple rural clothes, looking almost deliberately dowdy. They would rather have worn something smarter, but Khrushchev was most anxious to display them as humble peasant folk. Anna was in tears of pride, but Gagarin must have known how frightened she had felt over the last day or so. He hugged her, wiped away her tears with a handkerchief and said in a mock-childish voice, 'Please don't cry, Mamma. I won't do it again.'

The ceremony at Vnukovo was quite brief. The more important event of the day was in central Moscow. The Gagarins and the Khrushchevs boarded a black Zil limousine and headed for Red Square. Zoya thought her famous brother looked pretty much the same as usual, if rather tired and pressured.

That day Gagarin was assigned a rare privilege, a personal driver. Fyodor Dyemchuk collected from the authorities a brand-new Volga-21 car, complete with the latest and most fashionable accessory: a third foglamp. From now on, he and the Volga would be assigned permanently to Major Gagarin.

Sergei Korolev was not so well treated. He also met Gagarin at Vnukovo, but the Chief Designer was standing slightly to one side of the main reception group, and Khrushchev made no obvious move to acknowledge the man who, more than any other, had made this triumph possible. Korolev was not granted a bright, new Volga car. He bought an older Chaika limousine from one of the foreign embassies, so that he could at least get himself to Vnukovo in reasonable style, for no one else seemed much concerned to put him on display. He was a State secret. He could not be spoken of, let alone paraded in public. They would not even let him wear his medals. To cap it all, his second-hand Chaika broke down on the way to Moscow when the fan belt snapped, and he was forced to hitch a lift to Red Square in a more modest vehicle. In the long official list of scientists, soldiers, Academicians and politicians attending the celebration to mark man's first journey into space, Korolev's name does not appear. His colleague Sergei Belotserkovsky says today, 'The situation for Korolev was very unfair, and Yura was upset by that. The Nobel Prize Committee asked if they could make an award for the creator of the world's first satellite and the man who'd sent the first human into space, but the authorities never replied to them. Even today that injustice hasn't been remedied.'

In Red Square, Gagarin and his family stood alongside Khrushchev and the other Party leaders on the traditional perch of communist power: the reviewing stand atop Lenin's Mausoleum. Overhead, helicopters flew over the city's major thoroughfares dropping leaflets. The Red Army clumped and thumped across a cleared area of the square, but the greater allocation of space was given over to an immense cheering crowd. The façade of the GUM department store was obscured by a huge portrait of Lenin inscribed with the slogan, 'Forward to the Triumph of Communism!' Today, at least, that triumph seemed well within the bounds of possibility.

Not that much propaganda work, other than simply stating the facts, needed to be done to boost the day's glory. The Soviet Union had put a man into space. Nikita Khrushchev's senior aide and speechwriter Fyodor Burlatsky remembers, 'I was in tears, and many people in the streets were crying from the shock – a shock of happiness, first of all because a man was flying in heaven, in the realm of God, and most important, because he was a Russian. The mood of celebration was almost entirely spontaneous. Usually in Russia, during Stalin's time, and even during Khrushchev's time, these demonstrations of popular feeling were heavily orchestrated, but this one wasn't. It was natural, straight from the heart of maybe ninety per cent of people in the Soviet Union.'

Titov and several of the other cosmonauts from the first group attended the celebrations in plain clothes. They did not get to stand on top of Lenin's Mausoleum alongside Gagarin and the top brass, but had to stay at ground level. 'I saw this sea of people, not a sea but an avalanche of shouting, smiling people. They were lifting children onto their shoulders to let them have a look. Yura was standing on the Mausoleum among the government members,' says Titov. 'It was astonishing to see him there. It was only then I realized the importance of the event which had moved all the people. Everyone was glad. The whole world was glad because a man had gone into space. It was extraordinary.'

Gagarin made a speech from the podium: the usual sentiments that one would expect on such as occasion, but delivered with his own particular cheerfulness and sincerity, so that all the platitudes about communism, the Motherland, the Party, seemed for a moment genuinely to come alive. In conclusion Gagarin said, 'I should like to make a special mention of the fatherly love shown to us, the Soviet people, by Nikita Sergeyevich Khrushchev . . . You were the first to congratulate me warmly on the success of the flight a few minutes after I landed . . . Glory to the Communist Party of the Soviet Union lead by Nikita Sergeyevich Khrushchev!'

Lest any of his enemies be waiting in the wings for signs of weakness, Khrushchev flaunted his invincibility, his intimate connectedness with today's triumph. Gagarin's speech of gratitude – delivered with tact and sincerity to an immense and exultant crowd – was precisely what the First Secretary wanted to hear. The young cosmonaut became a firm political favourite from that moment. Bursting with pride and happiness and wiping tears of joy from his eyes, Khrushchev repeatedly hugged Gagarin, then made a hefty speech, to which the crowd listened with rapt attention, interrupting him at frequent intervals with long bursts of heartfelt applause.

One has to imagine that on April 14, 1961 the Soviet Union truly believed in itself, in no small part because Khrushchev understood how to win the loyalty of his people with an ebullient brand of showmanship. For now, at least, the blood-soaked struggle of the socialist revolution seemed to have taken flight under his more optimistic style of leadership. Joseph Stalin had instilled obedience on pain of death, but Khrushchev was immeasurably less terrifying in his desire for affection won without duress. On coming to power he had gone so far as to denounce Stalin's cruelties. This was a tremendous political risk, given that many surviving administrators from the old regime were still pulling the Party strings and did not want to be told that their travails under Stalinism had amounted to a terrible mistake; but today, with Gagarin's world-shattering achievement under his belt, Khrushchev was unassailable. For now.[10]

Of course the young man who had helped deliver him this wonderful victory would benefit from the First Secretary's warmest and most personal gratitude in the coming months and years. Unfortunately, winning Khrushchev as a friend also meant gaining his rivals as enemies. When Khrushchev's deputy Leonid Ilyich Brezhnev congratulated Gagarin at Vnukovo Airport and treated him as a lofty equal atop the Mausoleum in Red Square, he did so with all due comradeship and sincerity, but his body language – preserved in the documentary films of the event – betrays his lack of real warmth. In October 1964 his deference towards the First Cosmonaut would vanish overnight, along with Khrushchev's hold on power.

In the evening there was a celebratory dinner in the huge Georgyevsky Hall of the Kremlin. It was supposed to have been a luncheon, but the Red Square celebrations had lasted for a good six hours. The crowds' enthusiasm and proud patience seemed limitless, and Khrushchev had milked the day for all its glory.

At the dinner, a hungry and footsore Valentin tucked into all the food and booze with hearty enthusiasm. 'There was a huge round table – an entire delicatessen, I would say. Yura was awarded a Golden Star and the Order of Lenin. The last people to congratulate him were the holy fathers. There was one of ours, two Muslims, and a couple of others. One of them asked, "Yuri Alexeyevich, did you see Jesus Christ far up above the earth?" He replied, "Holy father, you'd know better than me whether I'd have seen him up there."'

Valentin noted to his satisfaction that sensible supplies of good vodka accompanied all the place settings, for the men at least. 'There was a bottle of Stolichnaya by everyone's side – not the modern stuff, but the kind you drink and then want to drink more of. There was also cognac, wine, and three glasses. I wasn't sure which glass to take, so I decided to copy Father. He took the middle one and I did the same. When I'd drunk from the glass, I asked, "How much are non-Party members allowed to drink?" Everyone went quiet. Only then did I realize my mistake. Father replied, "That's right, Valentin Alexeyevich. These days, Party members get to drink twice as much as non-members!" Everyone started to laugh and the tension disappeared.'

Valentin had another tot to recover from his embarrassment, and kept an amused eye on a group of Muslim delegates from the southern republics. 'You know, they were especially drunk. They have total abstinence, don't they? But here the drink was free of charge. They were great fun, and so were the Yugoslavs. The Poles also drank quite well. Some people were carried out by their elbows and put in their cars.'

Unfortunately, getting at the food was more of a problem than obtaining drink. 'There were no waitresses to serve us. So it was just like communism. You can sniff it and look at it, but you can't touch it or eat it. Furthermore, Khrushchev was shouting all the time that true universal communism was just on the horizon.'

Indeed, a triumphant Nikita Khrushchev was by now well into one of his noisy table-thumping routines. Cosmonaut Alexei Leonov recalls the mood of optimism. 'He announced that our generation was going to live in true communism. We were all hugging, applauding, screaming "Hooray!" And we really believed him, because at that time the success of our country was obvious to the whole world. It was only much later, when we grew up and learned a little about economic realities, that we realized Khrushchev's announcement was a little premature.'

*   *   *

By the end of April Gagarin's arduous routine of foreign travel was under way, with a trip to the 'democratic' satellite socialist countries of Czechoslovakia and Bulgaria, then on to Finland. In June 1961 he arrived back in Moscow for some much-needed leave with Valya and the children, although he took the time to be interviewed by yet more journalists, ferried from place to place as always by his driver Fyodor Dyemchuk. The Indian writer Khwaja Ahmad Abbas observed of Gagarin:

> He was widely hailed as the Man of the Moment, but when I came face-to-face with him the meeting began with an anticlimax. The door opened and the world's most publicized man stepped in, but I failed to recognize him. Even while shaking hands with him I was uncertain that this slightly built young man could be the great Hero of the Space Age. Even in his smart uniform he looked like a junior officer coming in as an advance guard to announce the real hero.[11]

As usual, Gagarin's charm quickly won the day and Abbas's approach soon became more complimentary. The journalist may have felt an initial twitch of disappointment, but the First Cosmonaut's essential normality was the whole point. If Khrushchev and his advisors had wanted a super-hero to represent the Soviet Union in space, they would have chosen another candidate.

The British journalists Wilfred Burchett and Anthony Purdy met Gagarin at the Foreign Correspondents' Club in Moscow on June 9, and were instantly impressed by his enthusiasm, his firm handshake and confident responses to their questions. They told him they were writing a book about his exploits, and he flattered them by saying that if their determination as authors was anything to go by, 'The next person in space should be a writer.'

The conversation soon turned to the exploits of NASA. Gagarin made gentle, but pointed, fun of the Mercury project, which had achieved only its first tentative 15-minute sub-orbital hop on May 5, with astronaut Alan Shepard aboard. Burchett and Purdy suggested that the American capsule incorporated more sophisticated attitude (orientation) controllers, thrusters and navigation systems, so that Shepard could genuinely pilot the machine to a greater extent than a cosmonaut could fly a Vostok. This was quite true, but Gagarin evaded the issue by concentrating on the short duration of the Mercury mission. 'How much driving can you get done in five minutes?' he challenged. 'And what would be the point of manual control? I could have guided Vostok, had I wanted to. There was a dual-control, but the manual option was not necessary or important.' For a pilot this was like saying that his job – his essential skill – was completely irrelevant, but at the time Gagarin could hardly have said anything different.[12]

The journalists changed tack and suggested that Mercury's cabin equipment was better than Vostok's. Again, this was largely true. Gagarin countered, 'It's difficult to compare them. Vostok's cabin is very big, and the thrust of its engines much greater. We went higher and faster for a much longer time.'

Burchett and Purdy asked him which had been the worst moment during his flight? 'The re-entry,' he replied without hesitation – then collected his thoughts for a moment and efficiently covered his tracks. 'But "worst" is a comparative word. There wasn't really any particular bad moment. Everything worked, everything was organized properly, nothing went wrong. It was a walk, really.'

Not surprisingly, Burchett and Purdy missed the nuance. Phil Clarke, a modern British expert on Russian space history, suggests that if the story of Vostok's retro-pack separation failure had leaked out in 1961, it would have caused a sensation, but Gagarin remained consistently skilful at listening to his own answers and guarding against errors.

As always, the most sensitive issue was his method of landing. In the wake of his homecoming celebrations in Moscow, Gagarin was pressed for answers by suspicious foreign journalists. On April 17 the London *Times* correspondent wrote:

No details have been given about the method of landing. Asked point-blank about this at the crowded press conference, Major Gagarin, more hesitantly than in his other replies, skated over the questions with his answer: 'Many techniques of landing have been developed in our country. One of them is the parachute technique. In this flight we employed the system where the pilot is in the cabin.' The pictures published in the press here also give little idea of the spaceship's structure, but some light was thrown on Major Gagarin's pride in it when he was seen to wince at the use of the word 'plane' at the press conference.

The sports official Ivan Borisenko flew to Paris in July 1961 to negotiate far more searching questions thrown at him by the International Astronautical Federation (IAF) about the altitude record he was claiming on behalf of Vostok. The IAF Director-General asked Borisenko's delegation outright, 'Where was the pilot on return, in relation to the space vehicle?' Borisenko bluffed shamelessly. 'Ask the Americans if they believe these records for Gagarin were actually achieved! All the people of the world have already endorsed Gagarin's flight and have accepted it as fact.'[13]

The wrangling went on for several hours, but eventually the IAF caved in without pressing the Soviets for clearer evidence. From now on Borisenko could wave his newly minted IAF certification in front of sceptics as 'proof' that Gagarin had landed in his ship and rightfully claimed the altitude record.

On July 11, 1961 Gagarin and his escorts flew to London on a Tupolev-104 Aeroflot airliner. The left-leaning London newspaper the *Daily Mirror* heralded his arrival with a glowing tribute, accompanied by a bitter critique of the lacklustre official reception. Today the piece can be read as an eerie portent of things to come, as the Conservative government of the time began to collapse under the pressure of 1960s' modernity:

Gagarin is a brave man, the symbol of one of the greatest scientific feats ever achieved. Yesterday, after two days of stuffed-shirt panic over the correct procedure, the British government at last figured out how they would welcome this world-wide hero. And who are they sending to greet

him in the name of the entire British people? Not the Prime
Minister, Mr Macmillan. Not the Foreign Secretary, Lord
Home. Not even the Minister of Science, Lord Hailsham.
Britain's spokesman on this unique occasion will be an
unknown civil servant, Mr Francis Fearon Turnbull, CBE,
aged fifty-six. The reason given . . . is that Gagarin is not
a Head of State.

Harold Macmillan did eventually meet Gagarin (though not at
the airport) and described him as 'a delightful fellow'. In fact,
Gagarin's visit to Britain had been sponsored largely by the
Foundry Workers' Union rather than the government, but the
ordinary citizens of Britain turned out in force to welcome
him. *The Times* reported that he 'received a welcome that
sometimes bordered on hysteria. Cheering crowds lined the
route into London all the way from the airport.' He arrived
by motorcade into the vast Earl's Court Exhibition Centre in
West London to address a crowd of students, then gave a press
conference in front of 2,000 journalists from Britain and around
the world. Quickly the establishment revised its plans for him. He
was summoned to the Admiralty, the Air Ministry and the Royal
Society, and finally to Buckingham Palace to meet the Queen.
Yaroslav Golovanov, the approved journalist ever in attendance,
says that an extra day had to be found in Gagarin's schedule
to make room for this meeting, which raises the fascinating
possibility that the royal reception was not planned in advance.
Rather, it was a hurried response to circumstances. *The Times*
seemed to confirm this on July 12, with its report that 'Because
of the Palace invitation, Major Gagarin will now return home
on Saturday, instead of on Friday as originally planned.'

During an informal luncheon encounter on July 15, the Queen
was gracious, particularly when Gagarin ran into the perennial
problem of an unversed visitor in the Palace at mealtimes: how
to handle the vast array of cutlery. Golovanov recalls the scene.
'He said, "Your Highness, you know this is the first time I've had
breakfast with the Queen of Great Britain, and it's very difficult
to know which cutlery to use." He smiled, and the Queen didn't

hold back. She said, "You know, I was born in this palace, but I still get mixed up." After that, the meeting went very warmly and sincerely.'

The Queen asked Gagarin all kinds of questions – simple human curiosity breaking through the pomp, as always – and at one point he said tactfully, 'Maybe you have me mixed up with someone else? I'm sure there are many other pilots like me in your own Royal Air Force.' In all, the First Cosmonaut was turning out to be an extraordinary asset to Soviet diplomacy, but, as he confessed to Golovanov in a quiet moment, the strain of playing the perfect ambassador was beginning to wear him down. 'A lot of articles are being written about the flight. Everyone is writing about me, and it makes me uncomfortable because they're making me out to be some kind of superhero. In fact, like everyone else, I've made mistakes. I have weaknesses. They shouldn't idealize people. It's embarrassing to be made to seem like such a good, sweet little boy. It's enough to make one sick.'

When he grew weary of the adulation at his news conferences, one of Gagarin's favourite ploys was to remind his listeners that his Hero of the Soviet Union medal was stamped with the number 11,175. 'That means 11,174 people accomplished something worthwhile before me. I disagree with any division of people into ordinary mortals or celebrities. I'm still an ordinary mortal. I haven't changed.' (Once, in Moscow, he laughed happily when he overheard a woman in the crowd say, 'Oh, look! He's cut himself shaving.')

On August 5 Gagarin's entourage arrived in Canada, at the invitation of the financier Cyrus Eaton. From Halifax they travelled 200 kilometres to Pugwash, Nova Scotia, where Eaton kept a substantial residence. He and the philosopher Bertrand Russell had convened a famous nuclear disarmament seminar known as the 'Pugwash Conference' at this house back in 1957. Not surprisingly, Moscow was delighted to receive an invitation for Gagarin to visit, but Eaton's star guest quickly became distracted. Late on the evening of August 8 he learned that Gherman Titov had gone into orbit. He asked if he could send a congratulatory telegram, and this was arranged for him by Nikolai Kamanin. Titov heard Gagarin's message during his sixth orbit, relayed to him by ground controllers. Cyrus Eaton politely eased a foreshortening of the festivities so that Gagarin and his colleagues could set off back to Russia immediately. All of a sudden the Russian delegation felt very cut off from important events at home. Kamanin remarked, 'While we're busy making speeches, the Americans are preparing spacecraft. We have to move ahead.'

Gagarin's tour resumed within three weeks. His arrival in Cuba on August 24 was a politically charged event, an important gesture of Soviet solidarity with Fidel Castro's two-year-old regime. Gagarin and Kamanin stepped off the plane into sweltering heat, dressed in dazzling white summer uniforms. Seen from their end of the political telescope, the Bay of Pigs had been a triumph, not a defeat. Castro's aides happily told Gagarin, 'The "beards" repelled the enemy,' and Gagarin replied, 'People who believe deeply in the rightfulness of their cause can never be brought to their knees.' As so often, he knew exactly what to say without prompting. At a mass public rally he declared, 'All two hundred and twenty million of us Soviet people are the true and devoted friends of Cuba!'[14]

By 1967, the last year of his life, Gagarin would not be so quick to praise the Soviet regime, or to take its every triumphant proclamation so much on trust.

8

THE SPACE RACE

Yuri Gagarin's short journey through space was one of the most important events of the twentieth century – not for Russia, but for America, where an industrial shake-up of colossal proportions was unleashed in response. It was not just Velcro fabric and the non-stick frying pan that emerged as a result of the Space Race, but the entire fabric of modern technology. Microchips were developed because 1950s' circuitry was not small enough to fit inside rockets and missiles. The Internet emerged from an attack-proof communications network laid down by ARPA, the Advanced Research Projects Agency (a pre-NASA government department that planned, among other things, for America's future in space). Modern diagnostic medicine owes an incalculable debt to the research conducted by the space doctors. The development of the global communications industry – for so long a science-fiction dream – happened with incredible speed after the invention of satellites. In all likelihood these technologies would have come along of their own accord, but probably nothing like as fast as they did. And all because a farmboy from Smolensk laid down a challenge to the most powerful nation on earth.

Dr John Logsdon, who heads the Space Policy Institute in Washington, DC, and has advised a succession of Presidents, explains the impact of Gagarin's flight on the American psyche. 'It was a sudden rebalancing of our power relationship with the Soviet Union, because of the clear demonstration that – if they wanted to – they could send a nuclear warhead across intercontinental distances, right into the heart of "Fortress America". There was an uproar: how did we get beaten by this supposedly backward country?'

President Kennedy had not taken space particularly seriously until now, but on the evening of April 14, 1961 he was deeply agitated at the global response to Gagarin's flight. He paced his office at the White House, asking his advisors, 'What can we do? How can we catch up?' Kennedy's science advisor Jerome Wiesner cautiously suggested a three-month study period to assess the situation, but the President wanted a more urgent response. 'If somebody can just tell me how to catch up. Let's find somebody – anybody. I don't care if it's the janitor over there, if he knows how.' He deliberately made these remarks within earshot of Hugh Sidey, a senior journalist from *Life* magazine. All of a sudden the President wanted to be seen as an advocate for space.[1]

Three days later Kennedy suffered another, more serious defeat. A 1,300-strong force of exiled Cubans, supported by the CIA, landed at the 'Bay of Pigs' in Cuba, with the intention of destroying Fidel Castro's communist regime. Kennedy had personally approved the scheme, but Castro's troops learned of the operation well ahead of time and were waiting on the beaches. The raid was a total disaster, because the CIA failed to deliver the promised support. Contrary to all expectations, the 'subjugated' population of Cuba showed absolutely no desire to participate in Castro's overthrow. To the CIA's lasting embarrassment, no attempt was made to rescue the invaders.

The Kennedy administration seemed to be faltering in its first 100 days, the traditional 'honeymoon' period during which a new president is supposed to shake things up and make his mark. Kennedy immediately turned to space as a means of reviving his credibility. In a pivotal memo of April 20, he asked Vice-President Lyndon Johnson to prepare a thorough survey of America's rocket effort:[2]

1. Do we have a chance of beating the Soviets by putting a laboratory in space, or by a trip around the moon, or by a rocket to land on the moon, or by a rocket to go to the moon and back with a man? Is there any other space program which promises dramatic results in which we could win?

2. How much additional would it cost?

3. Are we working 24 hours a day on existing programs, and if not, why not? If not, will you make recommendations to me as to how work can be speeded up.

4. In building large boosters should we put our emphasis on nuclear, chemical or liquid fuel, or a combination of these three?

5. Are we making maximum effort? Are we achieving necessary results?

This single-page document can be read either as one of the most sensational directives of the twentieth century or as a hastily dictated panic response to a bad week at the White House, but without doubt it laid the foundations for the largest technological endeavour since the wartime 'Manhattan' development of the atomic bomb: the Apollo lunar landing project.

NASA's chief administrator James Webb certainly believed that the Soviets could beat America at the short-term goals outlined in Kennedy's famous memo, such as orbital rendezvous and simple space stations. He suggested a landing on the moon as a longer-range goal, requiring such a tremendous input of resources and technical development that, in all likelihood, the Soviets could not match it. Webb persuaded Kennedy and Johnson to take the longer view, because the short-term battle for rocket supremacy was already lost.[3]

A final decision hinged on NASA literally getting their manned space programme off the ground. On May 5, just twenty-three days after Gagarin had flown, US astronaut Alan Shepard was launched atop a small Redstone booster. His flight was not a full orbit, merely a ballistic 'hop' of fifteen minutes' duration. In contrast to Vostok's orbital velocity of 25,000 kmph, Shepard's Mercury achieved only 8,300 kmph. Vostok girdled the globe, while the Mercury splashed down into the Atlantic just 510 kilometres from its launch site. But this cannonball flight was enough to prove NASA's basic capabilities.

Immediately in the wake of Shepard's successful flight, Webb made good use of the opportunity to strengthen NASA's position.

His budget advisors within the space agency suggested that he should keep the cost estimates for a moon project as low as possible, if he was to obtain presidential approval, but Webb did not agree. Instead, in one of the most talented administrative bluffs ever attempted by a US civil servant, he doubled the estimates and presented them to Kennedy with an absolutely straight face. It represented a colossal sum of money: upwards of $20 billion in 1960s' dollars, projected to be spread over the next eight years.

Stunned by the figures involved, Kennedy nevertheless decided to support Apollo. In a historic speech before Congress on May 25, 1961 he said, 'I believe that this nation should commit itself to achieving the goal, before this decade is out, of landing a man on the moon and returning him safely to the earth. No single space project in this period will be more impressive to mankind, or more important for the long-range exploration of space, and none will be so difficult or expensive to accomplish.'

Meanwhile, Webb and Johnson, canny southern politicians both, began to lay down a subtle and far-reaching network of aerospace contract pledges, construction schemes and political patronage to ensure funding approval for Apollo across forty states. Within four years, NASA's spending would command 5 per cent of the nation's entire annual federal budget, and would employ upwards of 250,000 people from coast to coast. Webb called his financial bluff an 'administrator's discount'. Modern NASA was built on his belief that the agency had only this one chance to establish itself as a permanent feature of national life, before the momentum for space exploration slackened off. He was quite right. Over the last four decades no other President since Kennedy has been so supportive, nor so willing to spend money on space.[4]

All this happened because Alan Shepard flew just twenty-three days later than Yuri Gagarin. John Logsdon raises a fascinating conjecture. 'The flight that Shepard made on May 5 1961, just three weeks after Gagarin, should have happened in March, but a previous Mercury test on January 31 had a chimpanzee called "Ham" on board. The retro-rockets fired late, sending

Ham [210 kilometres] downrange of the correct splashdown zone, and it took several hours to recover him – which made for one very unhappy chimpanzee, by the way. The technical problem was very simple, very easy to fix, but they had to do another test of the Mercury before committing a human. So it's an interesting question: what would have happened if Gagarin had been second? I think history would have worked out very differently.'

But Gagarin was first, and the American reaction was inevitable, particularly given the President's driven personality. Logsdon says, 'This wasn't a Soviet success, but an American failure. I don't think it was just a question of Kennedy's responding to public opinion [about Gagarin]. I think he had his own very personal reaction. He always had a very strong need to be first. He was a very competitive person . . . Perhaps he was looking for an opportunity to show leadership and take some kind of bold action.'

Hugo Young, a journalist from the London *Times*, observed something similar in 1969:

> Kennedy's response disclosed more than anything the sight of a man obsessed with failure. Gagarin's triumph pitilessly mocked the image of dynamism which he had offered the American people. It had to be avenged almost as much for his sake as for the nation's.[5]

Science advisor Jerome Wiesner surrendered to the inevitable, although he still could not see the point of spending such colossal sums on Apollo. He salved his conscience by forcing a promise from Kennedy. 'I told him the least he could do was never to refer publicly to the moon landing as a scientific enterprise, and he never did so.'[6]

The West quickly developed an obsession with the Space Race, much to the bemusement of cosmonaut Gherman Titov and his friends. 'What kind of race were they talking about? There wasn't a race, because we Russians were already ahead of the entire planet.'

* * *

Nikita Khrushchev and the Politburo did not immediately respond to Kennedy's speech with their own moon project. Instead, Khrushchev pressured Korolev for short-term rocket 'spectaculars', in the hope of demoralizing the American space effort before it became totally unstoppable. In the longer term, the Soviet economy could not hope to match America's staggering budgets for Apollo. However, as long as he continued to deliver results with Vostok and with his converted R-7 missile, Korolev could count on the Kremlin's continuing support for space exploration.

For the time being, NASA was also relying on converted weapon launchers rather than custom-made space boosters. On July 21, 1961 astronaut Virgil 'Gus' Grissom flew another sub-orbital curve in a Mercury capsule atop a Redstone ballistic missile, reaching an altitude of 190 kilometres. His mission nearly ended in disaster when the small hatchway on the capsule blew off shortly after he splashed down. He clambered out of the waterlogged craft without his helmet, and water poured down his neck ring and into his spacesuit. He tried to signal the approaching rescue helicopters for help and was amazed to see them flying over the capsule, instead of coming directly to his aid. The helicopter pilots thought Grissom was waving, not drowning. Ignoring him completely, they concentrated on trying to hoist the capsule out of the water before it sank, but it was now so heavy with water that it threatened to pull the helicopter down. Eventually Grissom was rescued, while his capsule sank to the bottom of the Atlantic, beyond all hope of recovery. NASA downplayed the fact that their astronaut was almost lost at sea and hailed Grissom's mission as a near-perfect success.

On August 6, Gherman Titov took off from Baikonur and flew seventeen orbits aboard the second manned Vostok, staying aloft for twenty-four hours and flying the ship manually for a short period. 'When I was launched, my wife [Tamara] went into the forest to pick some mushrooms. It was a Sunday, and she disappeared on purpose, to get away from all the journalists asking her annoying questions.' Titov was extremely nauseous during his flight, the heating in the cabin broke down so that he nearly froze, and his retro-pack did not separate cleanly before re-entry, which gave him some cause for concern: 'Whether you need this, eh?' His ejection and landing in the Saratov region were also hazardous, as he recalls today. 'Under my parachute I passed about fifty metres from a railway line, and I thought I was going to hit the train that was passing. Then, about five metres from the ground, a gust of wind turned me around so that I was moving backwards when I hit the ground, and I rolled over three times. The wind was brisk and it caught the parachute again, so I was dragged along the ground. When I opened my helmet, the rim of the faceplate was scooping up soil. You know, the farmers in Saratov had done their ploughing quite well that season, otherwise my landing would have been even harder.'

The train screeched to a stop, and a small crowd of people jumped out and ran towards Titov, who recalls that he was not in the best of moods by then. 'I said to them, "What are you staring at? Help me take off my spacesuit. I'm very tired." There was supposed to be a clean lightweight overall for me to get into, but as usual somebody forgot to pack it in my emergency kit.'

A civilian woman arrived by car, who she had been in such a hurry to drive into the field that she drove over a pothole and banged her head on the steering wheel. To his bemusement, Titov found himself applying the bandages from his emergency space medical kit to her wound.

Sick, exhausted, bruised, befuddled but alive, Titov was the first man to spend an entire day in space, and the first to make multiple orbits around the earth. Perhaps he can take additional satisfaction from beating Gagarin at his own game. Saratov is 1,500 kilometres away from Baikonur, which means that Gagarin's historic but incomplete first orbit on April 12 fell short by that distance; so Titov was actually the first man to complete a whole orbit. With all the fuss in the aerospace history books about Gagarin's 'altitude record', this detail seems to have slipped everyone's attention.

Titov is philosophical about the hazards that he and Gagarin endured aboard the temperamental Vostoks. 'I can't say I was ready for any of it, but we couldn't train for these malfunctions, because with so few flights behind us, nobody knew what kind of things could go wrong. Yuri and I came up with our own emergency manual [for Vostok] and we tried to anticipate all the things that could create difficulties. I can tell you, that manual was a pretty thin document. When you drive a car, you have to expect a puncture at some point. Machines in motion have a right to go wrong.'

One week after Titov's landing, construction of the Berlin Wall began. According to Korolev's biographer James Harford, Khrushchev ordered the timing of Titov's flight deliberately to encourage the German Democratic Republic (GDR) in its loyalty to Moscow.[7] However, there was more to the relationship between Khrushchev and Kennedy than simple-minded East–West aggression. It is now becoming clear that the two men considered the possibility of collaboration in space. John Logsdon, who has researched this issue, says, 'Kennedy was ambivalent about space even after he'd announced Apollo. In his Inaugural Address he suggested that the United States and the Soviet Union should explore space together, and that wasn't simply rhetoric. He created a group of advisors to look for ways of increasing cooperation . . . He sent feelers through his brother Bobby into back channels at the Kremlin . . . We're learning now that Khrushchev was ready to say, "Yes, let's explore, let's do it together." If those two men had survived, history might have been different, but Kennedy was replaced by Lyndon Johnson, a very strong nationalist, and Khrushchev was ousted by Brezhnev, who was far more militaristic . . . The contingency of history is that you end up with things as they happened, not as they might have been.'

NASA considered a third sub-orbital mission to qualify Mercury for more demanding missions, but the puny Redstone rocket's more powerful replacement, the Atlas (a fully fledged ICBM), was now ready to take a capsule into a full orbit, and the sub-orbital 'hop' schedule was cancelled. On February 20, 1962, John Glenn completed three full orbits and was applauded by US citizens, just as Gagarin had been in Russia. NASA also began testing unmanned prototypes for their giant Saturn lunar rockets. Admittedly they were only half the size of the superboosters that would eventually carry Apollo to the moon, but already they were close to outstripping the power of Korolev's R-7.

Korolev urgently needed to develop a successor for Vostok, which was limited in its capabilities, but Khrushchev wanted further triumphs within weeks and months, not years. Andrian Nikolayev was launched on August 11, 1962 for a four-day mission, and Pavel Popovich went up the very next day for a three-day stint. For the first time two people were in space simultaneously. Korolev timed the launches so that the second Vostok would briefly come within seven kilometres of the first: a cosmic clay-pigeon shoot, which enabled the Soviets to claim a 'space rendezvous'. In fact, the two craft quickly drifted apart and could never have regained their initial close formation. Their small rocket motors lay dormant throughout the flight, conserving fuel for re-entry braking at the end of the mission. However, appearances counted for a great deal. A number of aerospace professionals in the West were fooled into thinking that the Soviets had developed genuine rendezvous skills. In a 1995 interview with space historian James Harford, Vasily Mishin (who succeeded Korolev as chief of the OKB-1 Design Bureau in 1966) said, 'With all the secrecy we had in those days, we didn't tell the whole truth . . . As they say, a sleight of hand isn't exactly a fraud. It was more like, our competitors [in the West] deceived themselves all on their own. Of course we didn't want to shatter their illusions.'

The double Vostok mission seemed far in advance of American achievements. NASA launched astronaut Scott Carpenter on May 24, but it was hard not to see Mercury as a 'second-best' programme, by comparison with the remorseless originality of the Soviet missions. Even in the immediate wake of the Vostok double-flight, NASA's response was simply 'more of the same', with an October 3 mission for Walter Schirra lasting just nine hours.

A few days later, US spy planes photographed Soviet missiles in secret bases on Cuba, and public attention in that terrifying autumn of 1962 turned from space to the very real prospect of global nuclear war – not as some dim and distant prospect but as a horror that might be unleashed at any moment. President Kennedy initiated a naval blockade of all Soviet shipping approaching Cuba. Given the stark alternatives – a formal US invasion of Cuba or an air strike against the missile bases – the blockade seemed the least dangerous of several extremely high-risk strategies available to him. As is now clear from recently released tapes of White House crisis meetings, Kennedy and his staff went to bed on the night of October 23 not knowing if they, or anyone else in the world, would be alive next morning. Kennedy and Khrushchev very nearly backed themselves into a corner from which they could not withdraw.

Boris Chertok, a senior rocket engineer, recalls in an interview with James Harford that another attempt to launch a Mars probe from Baikonur that fateful October was interrupted when the military ordered Chertok 'to get the launch vehicle off the pad so they could replace it with an ICBM, because there was some kind of a national emergency. The military was using all the phone links, so I couldn't get hold of Korolev, who was at home in Moscow ill with a cold. They told me I'd be court-marshalled if I didn't move the Mars rocket away from the pad, and they began to check the systems on their missile. Only Korolev could have got through to Khrushchev to countermand these terrible instructions.'[8]

Chertok flew to Moscow and went to Korolev's house, where the Chief Designer solved the problem with a quick call to the Kremlin. By a horrible irony, when the Mars probe was launched on October 24, right in the middle of the Cuban crisis, it blew up, causing the ultra-alert US Ballistic Missile Early Warning System (BMEWS) to suspect a nuclear attack. Fortunately, the BMEWS tracking computers got to grips with the situation within a few seconds, and a counterstrike was not initiated.

Another terrible irony: the catastrophic R-16 missile explosion at Baikonur on October 24, 1960, which killed 190 people, may have precipitated the Cuban Missile Crisis. Quite apart from the vainglorious Marshal Nedelin, a significant number of skilled military missile engineers died that day, and the development of a credible intercontinental Soviet nuclear-strike capability was severely delayed by their loss. For the time being, US strategic targets could not be reached from within the Soviet Union. The one fully functioning ICBM available to Khrushchev was the Chief Designer's R-7, which took too long to prepare for launch and was not available in sufficient numbers to pose much of a threat. The missiles despatched to Cuba were simpler and less powerful battlefield nuclear weapons: small, numerous, easy to launch, but capable only of short flights.

Once the Cuban crisis was over, Khrushchev may have thought it better to carry on playing harmless space games with the Americans rather than risk all-out nuclear confrontation on the ground. As far as can be judged, the next orbital propaganda coup seems mainly to have been his idea. He wanted Korolev to launch the kind of person no one had thought of before: a woman.

By 1962, up to 400 women candidates had been screened for a possible space flight. On February 16, five were selected for training. In keeping with Khrushchev's orders, the women were chosen from the ranks of peasants and factory workers, rather than from specialist scientific or academic professions. The most suitable candidates were those who could combine humble origins with at least some kind of suitability for space flight. It just so happened that parachuting was becoming a popular hobby for many ordinary women in those days. The most favoured candidate to emerge for space training was Valentina Tereshkova, a 25-year-old parachuting champion with fifty-eight jumps to her credit. Tereshkova's father, a tractor driver on a collective farm, was killed in action during the war. Her mother became a weaver in a textile plant, and Valentina herself learned a similar trade. She was ideal for Khrushchev's purposes: fit, handsome, sufficiently smart for the intellectual challenges of space training, but not so advanced in her education that she could not safely represent the ordinary peasant and working classes.[9]

Gherman Titov remembers how the female cosmonauts were

greeted with distrust when they arrived at Star City. 'Frankly speaking, we didn't believe that womenfolk belonged anywhere near a flying machine. At that time, we thought that only men could carry out all the tasks involved in space flight. When the first women arrived, my feelings were negative. As everyone knows, Titov states his opinions on everything! But in the end we saw it was quite correct to have female cosmonauts, and we soon thought of them as good fellows, just like us.'

At last Yuri Gagarin was reassigned to a meaningful work schedule. At Star City he was appointed to head the female cosmonauts' training programme, in conjunction with his cosmonaut colleague and friend Andrian Nikolayev. On July 12, 1962 he was promoted to Lieutenant-Colonel, and eased his way back into the rhythm of real work by serving as prime radio communicator (what NASA would call a 'CapCom') during Nikolayev and Popov's double Vostok flight in August. Then he laid down an intensive physical training regime for his five female students (though he was still called away at intervals for yet more foreign trips).

By now, the training for a fully automated Vostok mission was not particularly difficult. Tereshkova's eventual flight on June 16, 1963 presented very few fresh technical challenges. She made a close approach to another Vostok carrying Valeri Bykovsky, but, as with the previous double-flight, the real trick was in the timing of the respective launches. However, her success gave Khrushchev the opportunity he was waiting for – to gloat about 'the equality of men and women in our country'. Once this propagandist experiment was successfully completed, the women's cosmonaut squad was quietly disbanded. On November 3, 1963 Tereshkova and Nikolayev were married in Moscow in a very public ceremony. It was the social event of the season, much enjoyed by Khrushchev. Three days later Gagarin was promoted to full Colonel. It seemed as if his career was progressing, but, as he would discover, higher rank could hinder as well as help him in his quest for another flight into space.

Bykovsky and Tereshkova's double-flight was the last mission for Vostok in its current configuration. By the summer of 1963 there was no longer any need for the Soviets to compete with

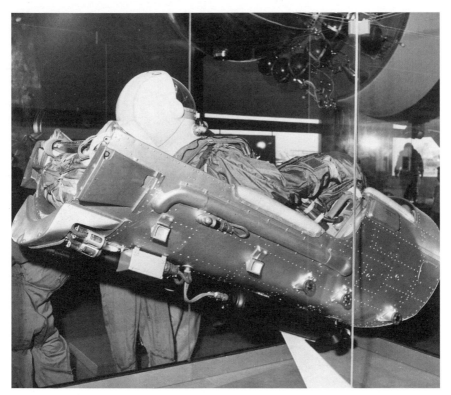

A Vostok ejection seat and space suit

Gagarin and his political champion, Nikita Khruschev, celebrating in Moscow, 14 April 1961

Gagarin with Sergei Korolev shortly after the flight, 1961

Breakfast with Nikolai Kamanin, summer 1961

Autographed
photo of Gagarin
with Fyodor
Diumtchuk, his
personal driver

Fidel Castro welcomes the First Cosmonaut to Cuba, August 1961

Veniamin Russayev,
Gagarin's KGB
escort and good
friend, with
Valentina

A kiss from one star to another: actress Gina Lollobrigida fulfils a
personal ambition

A present from France: Gagarin's pride and joy, a Matra sports car

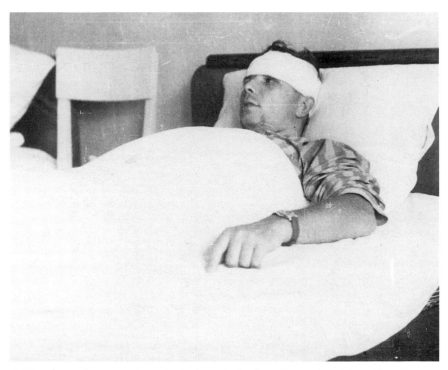

Fallen from favour: Gagarin in hospital after disgracing himself at Foros

The pressures of fame begin to tell on Yuri and Valentina during their Japanese tour, May 1962

Gagarin's hard partying worried his superiors

Konstantin Feoktistov, Boris Yegorov and Vladimir Komarov after their
mission aboard Voskhod I, October 1964

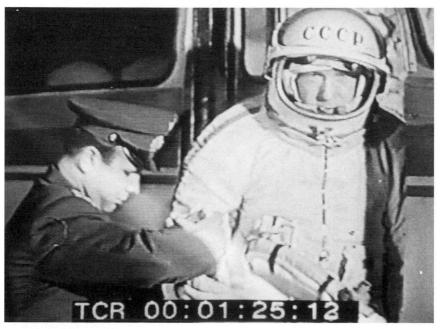

Gagarin escorts Leonov to the pad prior to the first walk in space,
March 1965

Banned from space in 1967, Gagarin makes a fatal return to flying jets

America's Mercury. After launching Gordon Cooper for a 34-hour mission on May 15, 1963, NASA decided that the programme had nothing else to prove. What more could these simple one-man capsules possibly accomplish?

In fact, NASA was planning to use a new missile, the 'Titan', developed by the US Air Force and reluctantly hired out on licence. Titan's fuel tank and outer skin were one and the same component, which saved weight. On the launch pad, the rocket was so flimsy that it could only stand upright if it was pressurized with inert gas, but it was so powerful in relation to its weight that it could carry a payload much heavier than the Mercury capsule.

NASA knew that its three-man Apollo moonship and its huge Saturn V rocket were years away from a first flight. So far they had only designers' mock-ups, not real hardware, to play with. Meanwhile, an interim vehicle was developed: a cross between the simplicity of Mercury and the complexity of the emerging Apollo. 'Gemini' was a two-man capsule designed to mate with the Titan missile. It incorporated ejection seats and hatches that could be opened in orbit to allow for spacewalks. The design was familiar to Korolev from NASA's open literature.

By 1963 the Gemini capsules were under construction at the McDonnell Douglas plant in California, but none of them had actually flown yet. Korolev was anxious to start work on a successor to Vostok, a larger capsule to match or even surpass NASA's new design. If he could launch what appeared to be a multi-man craft before the first Gemini was launched, then he would gain political support for building a genuinely more powerful competitor. Certainly Khrushchev wanted him to conjure up a three-man mission as soon as possible, to trump Gemini and embarrass the Apollo effort, although the extent to which he supported the taking of risks with cosmonauts' lives to achieve this goal cannot be judged today. Khrushchev is often blamed for pushing Korolev into hazardous decisions, but he could not possibly have decided all the technical details. He must have trusted the Chief Designer's judgement as to whether or not a particular space project was safe.

Korolev took a risk. He decided to adapt the current Vostok

hardware to carry two cosmonauts in the same ball – and even three, if they gave up their spacesuits. This new seating arrangement was purely cosmetic; it did not make the Vostok any better, just more cramped and significantly more dangerous. The bulky ejection seats had to be sacrificed in order to make room for the extra men, and if anything went wrong on the launch pad, there was no chance of escape. The new scheme was called 'Voskhod' ('Sunrise').

Despite its dangers, Voskhod would eventually prove capable of maintaining the Soviet lead, thereby adding impetus to Apollo, and also to Korolev's ambitious plans for his own moon shot. Vasily Mishin, his eventual successor at OKB-1, insists that the Chief Designer made a man-to-man deal with Khrushchev. Korolev would develop a multi-crew programme at very short notice in return for Khrushchev's approval for a giant new rocket, the N-1, a superbooster almost the equivalent of NASA's Saturn V.

By 1964 the space effort was securely established on both sides of the superpower divide. In August the Politburo approved development of the N-1, as well as two competing projects generated by Korolev's rivals in other sectors of the Soviet space industry. Ultimately this confusion, and Korolev's early death in 1966, would doom the Soviet moon programme to failure, but in the summer of 1964 almost all the cosmonauts were gearing up for ever more elaborate flights, with real hopes of planting their feet in lunar soil. To his dismay, Yuri Alexeyevich Gagarin found that he was no longer qualified to join in the fun.

It was not merely that his public duties were taking him away from his real work at Star City. Back in 1961 he had done something very foolish, only a few months after his historic flight, when he took a holiday and fell from grace.

9

THE FOROS INCIDENT

Crimea is almost an island. It juts out into the Black Sea, connected to the Ukraine by two peninsulas as delicate as veins. The northernmost territories of the island are pleasant but dull. The south is a different matter. There are beautiful mountains, sun-dappled forests, sheltered beaches speckled with palms. The weather is still fine in October, and the almond trees are back in bloom by February.

The surrounding Black Sea has never been quite so private a lake as Moscow might have liked. The southern half belongs to an old enemy, Turkey. Russia bears a grudge: at Balaclava in the Crimea, Lord Cardigan's Light Brigade charged into the Valley of Death, cannons to the left of them, cannons to the right of them . . . but Russia eventually lost that war, in part because of the Turkish contribution. From Sevastopol, the Black Sea Fleet's rusting hulks still maintain a wary watch on Turkey and its NATO allies.

At the Crimean port of Yalta, Churchill and Roosevelt made their uneasy wartime accommodation with good old 'Uncle Joe' Stalin; and in Foros, just along the coast to the west of Yalta, Nikita Khrushchev kept his *dacha*. Modern leaders still spend their summers here, though they can never be sure what their enemies might be planning while they are relaxing the best part of a thousand kilometres from the Kremlin. In August 1991 Mikhail Gorbachev was caught napping at his *dacha* high up on the cliffs, from where his view of the horizon obviously was not quite clear enough.

In its 1960s heyday, the Kissely *dacha* at Foros was a luxury sanatorium complex designed to accommodate only the most privileged group bookings. Warm seas, fresh meat and fruit, fine wines, perhaps a certain freedom from everyday restraints: all of these pleasures were available, and more. It was not expected that officials would record too closely how the guests at Foros enjoyed themselves.

The first cosmonauts and their associates also came to Foros for their holidays, with their wives and families in tow.

Call her 'Anna'; perhaps there were two Annas. Anna Ruman-seyeva, a young nurse, was on duty at the Kissely Sanatorium on September 14, 1961, when Gagarin and his cosmonaut comrades came to stay. She speaks with intimate knowledge of another nurse called Anna, also working at Foros when Gagarin came to stay. Maybe the two Annas are one and the same person? It is not important. Today Anna Rumanseyeva is a married woman, a respectable grandmother and professional medical practitioner.

'There are some people in life, especially men, who are constantly looking for adventure,' she says. 'I would say, Yuri Alexeyevich Gagarin was this kind of person. There was a small episode, a jump from a terrace – we can tell a short version of the story, yes? – I don't think he wanted to hide anything from his wife, Valentina. No, he was simply showing off, being childish, just to say to her, "You were mistaken, thinking I was in there, doing something wrong."'

The longer version of Anna's story is more revealing.

There were twenty-eight people in the group. Yuri and Valentina arrived at the sanatorium with their second daughter Galya, nine months old and still in need of her mother's constant attention. Gherman Titov was there; Alexei Leonov; the journalist Yaroslav Golovanov; a large crowd of cosmonauts; some technical people; even the dreaded Nikolai Kamanin, sharing a friendly drink with the boys, taking a rest from being (as Golovanov puts it) a 'complete Stalinist bastard'.

Kamanin noticed that Yuri and Valya were not getting along. He was rude, distracted and paid her very little attention. She would sit sulking in the car while her husband strode off to see the sights or meet with local Crimean dignitaries for a drink.

Sometimes he behaved so unpleasantly that Valya burst into tears. Kamanin and his wife Maria were shocked and surprised at Gagarin's behaviour. A few days into the vacation, Kamanin took him aside. As he noted in his ever-vigilant diary, 'I said to him, "This is the first time I've ever felt ashamed for you. You've offended Valya deeply." Gagarin admitted he was at fault and promised to mend his ways.'[1]

Titov's behaviour at Foros was hardly any better. The discipline so much admired by Kamanin in the lead-up to the first Vostok flight seemed largely to have evaporated by now. Kamanin felt the need to warn both his prime cosmonauts that they were 'slipping onto a dangerous path'.

Gagarin's conduct did not improve, and he seemed desperate for distraction. In the second week he took some of his companions out to sea in a small motor boat. The Foros staff pleaded with him: it was against the rules, he did not know the local conditions, the wind was offshore, the weather could be difficult, he should not go. But he went anyway, taking the boat far from shore and driving it recklessly, making tight turns to splash his passengers with spray. The swell picked up, just as he had been warned.[2] The boat was carried over the horizon and out of sight of the shore, and a larger motorboat had to be sent out to make a rescue. When they hauled him back ashore, Gagarin went to the medical station for assistance. In the rough conditions he had turned the boat's steering wheel so hard that his hands were bloodied and cracked. But the pain, and the unpleasantness of his foolish adventure, did not entirely divert his attention from the pretty blonde-haired nurse who attended to his blisters. 'Yuri Alexeyevich was a very nice person, merry and cheerful,' Anna admits. He asked whether she worked there? 'Yes,' she said.

The next day Titov, Kamanin and ten others of the group packed their kit bags for departure first thing the following morning. Of course, on their last day of freedom from care they celebrated hard. 'Then, in the evening, they celebrated some more,' Anna recalls drily. Kamanin described quiet games of cards and chess in his diary entry for that day, but this well-behaved tableau seems improbable, given the general pace of drinking and rowdiness established over the preceding fortnight.

The journalist Golovanov's version of events on October 3 is that 'Gagarin was a guest of the sailors in the Black Sea Fleet in Sevastopol. I was with him and Gherman Titov. Then we returned to Foros, and the next day we set off to the local Pioneers [the Russian equivalent of boy scouts] near Yalta. Then we visited the vineyards at Massandra. Basically we came back from there quite warmed up. Yuri decided to visit a lady friend. But we have to say something about his good character . . .' Golovanov re-directs the thrust of his story for a moment. 'You know, his wife Valentina was quite a complicated woman. She protected Yuri from every kind of temptation which came as a result of his position . . . Anyway, Valentina discovered that the First Cosmonaut had disappeared, and she decided to find out where he was, and he showed the true colours of goodness and of a gentleman. He showed genuine nobility and jumped out of a window on the second floor.'

Anna Rumanseyeva recalls, 'To avoid watching them playing and joking at the party, "Anna" had to leave the building. She said she went into a room and sat on a sofa. Yuri Alexeyevich – I don't know what was on his mind. He was drunk. Perhaps he wanted to talk? I don't think he had any other thoughts. Anyway he went into the room. He closed the door but didn't lock it with a key. Valentina Ivanovna went into the room immediately after him. The door opens . . . Perhaps he wanted to say that she was mistaken, or perhaps he wanted to hide? I don't know.'

After the incident Nikolai Kamanin interrogated various members of the sanatorium staff, including Anna, and decided on his own version of events:

Nurse Anna told me she had just gone up to her room to have a little rest at the end of her shift. She was lying on the bed fully dressed, reading a book, when Gagarin came into the room, locked the door behind him and tried to kiss her, saying, 'What, are you going to cry for help?' There was a knock on the door at that moment, and Gagarin jumped from the balcony.

Perhaps there was an unpleasant discussion between Yuri and Valentina, or perhaps she burst into the room and found no sign of him, just a breathless and dishevelled Anna. Perhaps Valentina demanded to know where her husband was, so that Anna had to tell her he was hiding on the balcony. Anna's accounts of the scene are many and varied – necessary interpretations rather than outright falsehoods – but of course both women leaned over the balcony's edge to take a look, as they had to, and saw Gagarin sprawled on the ground, motionless. 'At that time, there were wild grapes growing on the balconies,' Anna Rumanseyeva explains. 'They may have caught him as he jumped. He hit a kerbstone with his forehead. It was not a good landing. On his return from space he landed successfully. Here, unsuccessfully . . . I learned this from "Anna". Her name was also Anna. She told me.'

Nikolai Kamanin's first reference to the incident in his diary is brief and to the point:

> Under alcoholic intoxication, Gagarin jumped out of a window. It caused serious trauma to his face and a scar above his eyebrow. An operation was performed by naval doctors. He stayed in hospital for more than a month and missed the Communist Party Congress.[3]

Kamanin was among the first to reach Gagarin where he had fallen. He was not best pleased at the cosmonaut's condition. There was so much blood that he imagined for a moment that Gagarin must have shot himself. Meanwhile Valya had run downstairs to see what had happened. She screamed at Kamanin, 'Don't just stand there! Help him! He's dying!'

Immediately, doctors from a field station at Sevastopol were summoned. Meanwhile, the Foros medical personnel provided some basic first-aid; they checked for feeling in Gagarin's limbs, then decided it was safe to put him on a folding cot, which someone brought to the scene from indoors. Then they took him inside, where the doctors applied local anaesthetic to his brow. Some of the bone in his forehead was chipped. When the Sevastopol surgeons arrived, they cleared out the fragments, effected temporary repairs and stitched the wound. Gagarin held someone's hand throughout. He made no sound whatsoever, but his nails left livid marks, so tight was his grip.

The enormity of Gagarin's blunder seemed to catch up with him. He looked up at the nurse Anna for a moment and she remembers him asking her just one question. 'Will I fly again?'

She said, 'We'll see.'

Anna Rumanseyeva was grateful that Gagarin took the trouble even now, in his pain and discomfort, to protect 'Anna' from the authorities. 'He asked for one of the sanatorium directors, and he said, "Of course you know it wasn't her fault." And it was so. She was moved to a different building, but she continued to work in the sanatorium.'

A special private medical facility was established in the main wing of the sanatorium. Anna and another nurse alternated their duty rota, keeping Gagarin under permanent observation, while Valentina spent many hours at his bedside. All things considered, she was remarkably friendly towards Anna. 'She recalled how they lived before Yuri went into space. She explained how he studied hard, and she did regret that life sometimes.'

At the scene of Gagarin's accident, the doctors feared that he might have sustained concussion injuries. Afterwards, Yuri insisted that he had never actually lost consciousness, but a strict regime of bedrest was ordered nevertheless. After three days of inertia, he was propped up on his pillows, complaining to Anna, 'I'm fed up. I want to do something. Anna, please close the door. I want to do some hand-stands.'

'Yuri Alexeyevich! If the doctors find out, I'll lose my job!'

'Don't worry. I'm feeling healthy. I just want to do something.' He stood on his hands, larking about and feeling fine, but bored to hell. Anna persuaded him to get back into bed. He said, 'People will talk about this for the next hundred years. One day, when you're a grandmother, you can tell your grandchildren how you once took care of Yuri Gagarin.'

But he knew he had done a foolish thing in jumping from the window. Perhaps this adventure was unlucky for him. Behind the jokey smile and his irrepressible self-confidence, Gagarin brooded about his future.

Nikolai Kamanin was also concerned. He was responsible for maintaining discipline among the cosmonauts. In his diary he noted:

> This incident could bring a lot of trouble to me and others responsible for Gagarin. It could have had a very gloomy outcome. Gagarin was a hair's breadth from a very nonsensical and silly death.

Three days later a Chaika limousine arrived to take Gagarin to the Party Congress. He was carried on a stretcher, although he was up and about by now, and found the whole process absurd, laughing out loud. They took him to Sevastopol and put him on an aircraft to Moscow. On arrival he was not permitted to speak for too long at the Congress, or to mingle afterwards with the other delegates. The official records tell of his fully active participation, despite Kamanin's conviction in his diary entry that Gagarin was in no fit state to attend and did not take part. Golovanov explains, 'Actually he did turn up, but only on the fifth or sixth day after the opening of the session, and the photographer kept taking pictures of his profile so that the wound on his brow wouldn't be seen.' Meanwhile the newspapers put out a story to deter the curious. 'I remember they said Yuri was holding his baby daughter when he tripped, and so that the baby didn't get hurt, he sacrificed himself and hurt his brow. That's how they explained the wound.' In another version for *Izvestia*, Gagarin dived into the Black Sea to save his baby girl from drowning and banged his head on some rocks.

The doctors who had treated Gagarin were awarded commendations and promotions. Nikita Khrushchev was annoyed that his favourite cosmonaut could not give a proper performance at the Party Congress, but more than that, he was concerned for his young friend's safety. The moral aspects of the drama at Foros did not seem to concern him particularly. Khrushchev's advisor, Fyodor Burlatsky, says, 'In spite of the Party morality, which was supposed to be very strong, everybody thought it was a funny story. Khrushchev laughed. Maybe his wife didn't . . . But I think there were some Generals, high-level military people, who didn't have such easy relations with Gagarin. I think they were jealous because he was so close to Khrushchev.' These resentful rivals did not find the story quite so amusing, Burlatsky suggests. They noted Gagarin's behaviour with distaste, and remembered it.

Of course Nikolai Kamanin was severely criticized for his failure of supervision at Foros. At a special meeting in Star City on November 14 he had to explain Titov's and Gagarin's behaviour in the best possible light:

> Gagarin and Titov described their behaviour in the health resort adequately on the whole. They acknowledged their alcohol abuse, thoughtless attitude to women and other faults. However, probably for Valya's peace of mind, Gagarin maintained that he did not know the girl was in the room from which he jumped.

Kamanin persuaded the meeting to judge that his adulterous First Cosmonaut had merely been teasing his wife in a childish hide-and-seek game, while his drunken understudy Titov had been led astray by non-cosmonaut companions. Everyone knew that these were white lies, but tidy official versions were agreed, with the help of hand-written notes of apology from the cosmonauts themselves. Kamanin noted, 'I'm sure Gagarin had a different motive for visiting that room, but I won't press the matter, in case it causes discord in his family.'

Kamanin was lenient, but during the December resumption of Gagarin's world tour, he found to his great frustration that the First Cosmonaut's behaviour still was not improving. On December 14 he wrote in his diary:

He hasn't given up drinking, even after the Crimean incident. I don't fancy being a prophet of doom, but it seems to me he's drinking a good deal. He's at the top of his glory, bearing a great moral burden, knowing that his every step is being watched. One or two years will pass, the situation will change drastically, and he will become dissatisfied. It's obvious in his family life even now. He has no respect for his wife, he humiliates her sometimes, and she doesn't have the advantages of education or the social skills to influence him.

He also observed that 'Titov, recently returned from his tour of Indonesia, is starting to think no small beer of himself.' Evidently Kamanin felt that he had another wayward cosmonaut on his hands. But one has to keep in mind that his personal diaries are private expressions of annoyance, as much as accurate historical records. There is scarcely a single person within the Soviet space effort (not even the great Chief Designer) whom he does not criticize at some point – often unfairly – and just as often prior to a complete reversal of opinion a few days later. Even Khrushchev comes in for flak. Perhaps Kamanin's administrative tensions after the Foros incident can explain an extraordinary outburst of contempt in his diary about the Party Congress – the one at which Gagarin's muted appearance had caused so much embarrassment. At the Congress Nikita Khrushchev had proposed that Joseph Stalin's body be removed from the Mausoleum in Red Square. On November 5, 1961 Kamanin raged:

Many people disapprove of this. They speak about it openly in buses, on the metro and on the streets. The destruction of Stalin's prestige creates many problems. The young are losing their faith in authority . . . Stalin ruled the country for thirty years and turned it into a mighty state. His name can never be eclipsed by the pathetic pretensions of pygmies. Khrushchev is an envious intriguer, a cowardly toady. Everyone knows about his total diplomatic failures with China, Albania, the USA, France, England, and so on.

The irony is that no one in Stalin's time would have dared put such words on paper, for fear of being found out and shot. It can be assumed that Khrushchev's officials blamed Kamanin for not keeping his cosmonauts in order, so he vented his feelings in the pages of his diary; but he was not alone in his political opinions. It is hard for Westerners to understand the extent to which Stalin's memory was revered. Kamanin may have had a variety of reasons for his bitterness in October 1961, but he was broadly correct in his assessment. First Secretary Nikita Khrushchev was heading for a fall, and so was Yuri Gagarin.

By December 1961 the world tour had resumed, with Gagarin's slight scar carefully disguised by make-up. Delhi, Lucknow, Bombay, Calcutta, Colombo, Kabul, Cairo . . . Onwards, ever onwards. During the Ceylon visit, Gagarin carried out fifteen separate speaking engagements in one day. In Cairo a newspaper announced that he was nominated for election to the Supreme Soviet as a representative of the Smolensk region.[4]

Athens, Nicosia, Tokyo – and a loaded question about toys. A Japanese journalist wanted to know why Gagarin had bought a load of Japanese stuffed toys for his children. Could it be that Russian toys were not available back home? Gagarin replied, 'I always bring presents back for my daughters. I wanted to surprise them this time with Japanese dolls, but now this story will be all over the newspapers and it'll take away their surprise. You've spoilt a joy for two small girls.' He made this speech with the most charming smile and the questioner conceded defeat. The other journalists in the room buzzed their approval. Game point to Gagarin.

Valya came along on this leg of the tour, but it was not easy to combine foreign trips with childcare. She preferred to stay at home in Moscow while her husband travelled. She was shy, and found her occasional public appearances very difficult. This was not the life she had expected.

Fyodor Dyemchuk drove Valya around while Gagarin was abroad. He could hardly fail to notice her intense dislike of publicity. If her occasional foreign trips were a strain, then the streets and shops of Moscow were no less of a burden to her. Dyemchuk escorted her during household shopping trips, and Valya would always take her place in the inevitable queues like an ordinary Muscovite, but the other women in the line usually recognized her straight away. 'She would immediately turn around, get back in the car and say, "Let's go. They recognized me." Everyone would tell her to come to the front of the queue, but she would modestly come back to the car and go to a different shop.'

Nikolai Kamanin accompanied Gagarin on several more foreign trips. On December 4, 1961, during the visit to India, he wrote in his diary:

> Thousands of people greeted Gagarin warmly. I was reminded of my naïve childhood impressions of Christ meeting his people. He needed a miracle with five thousand loaves and fishes, but our Gagarin satisfies the people's thirst by his appearance alone. I'm the one writing these words, although I know better than anyone that Gagarin happens to be here only by sheer luck. His place could easily have been occupied by someone else. I remember writing on April 11 [the day before the space flight], 'Tomorrow Gagarin will become famous worldwide', but I could never have predicted the scale of his glamour.

By December 9, Gagarin, Valya and the accompanying entourage were in Colombo, Ceylon. Gagarin told Kamanin that he was 'close to wearing out'. The Soviet ambassador in Colombo insisted on his making as many appearances as possible. Kamanin could not help but note:

> They are doing their best to squeeze the maximum possible use out of Gagarin to make the government look good. They have no interest in how all this affects him.

By now Kamanin was growing concerned about Gagarin's alcohol abuse and Valya's increasing inability to deal with the stress of public appearances. Kamanin, Golovanov and other close colleagues have a similar view about this. It seems that Gagarin was a sensible drinker, a fun-loving man who could get drunk with the best of them but seldom drank too much when working. Unfortunately the publicity tours placed him in social situations where he was expected to drink each and every time, so as not to snub the endless toasts made in his honour. This, coupled with the emotional strain of his remorseless public schedule, inadvertently led him towards excessive drinking. Gagarin's personal KGB escorts and speech advisors, Venyamin Russayev and Alexei Belikov, were criticized for allowing this state of affairs to occur, although there was little they could do to prevent it.

Gagarin developed a very good relationship with Russayev, and each man worked behind the scenes to protect the other. Today Russayev says, 'Yuri was a very pure-hearted man, often taking responsibility for problems caused by others. As for Gherman Titov, problems seemed to slide past him like water off a duck's back. After his flight he had 20 or more serious disciplinary incidents, car crashes or whatever. People always wanted to connect these problems to Yuri, and at that point I would intervene.'

Russayev (and his colleague Belikov, an excellent linguist) sat close to Gagarin while he made presentations or spoke to journalists. 'I wasn't like the bodyguards; I was more an advisor and assistant. You have to imagine all the difficulties Yuri came across in his public life, and on the foreign trips. It was my job to look after him.' Not that Gagarin was socially inept – far from it. 'He had an excellent memory for the names of all the politicians and officials he met, quite unlike Kamanin, who came on many of the trips and was hopeless at that kind of thing . . . Often Yuri could play it by ear, and didn't need guidance. I was amazed at how he could cope with so many difficult questions.'

Russayev tells a touching story. 'Nikita Sergeyevich Khrushchev could get pretty tipsy on just a couple of glasses of booze, and Yura always protected him from going over the top.' Although Gagarin quite obviously adored Khrushchev, he kept as far apart from other politicians as he could safely manage without giving offence. Sergei Belotserkovsky came to know Gagarin very well, tutoring him in aerospace studies from 1964 onwards, when Gagarin was trying to get back to serious work. 'I think his personality began to split. On one side he was the welcome guest of kings, presidents and even the Queen of England, but on the other side he never lost his ties with the ordinary people. I think he began to sense the lower classes' lack of rights, their hardship, and he saw the corruption of the top layers of society. He saw our drunken leaders dancing on the table and behaving badly, and that can't have left his honest soul unwounded. I'm talking not just about the external symptoms, but also the internal corruption which was dominant among our top leadership clan.'

Quite unfairly, there was a certain degree of resentment directed at Gagarin because of all the privileges he was assumed to have accumulated under Khrushchev. True, he and the other cosmonauts were given better-than-average living accommodation, but their level of comfort was not significantly better than that of most middle-ranking officers. Titov says, 'Honestly, we never received special benefits. People were always saying to me, "Show us, point to the place where you'd like your luxury *dacha* to be, and Khrushchev will build it for you straight away." I didn't bother, and neither did Yuri. We were just young men. What were they talking about? What did we want with *dachas*?'

Russayev confirms this. 'Yuri was a completely honest guy. He was the first cosmonaut in space, he did so much for his country, and you should see the place that Valya lives in today. Instead of a decent *dacha*, it's a hen-house. Yuri worked hard for the good of his native land, not for his own wealth.'[5]

Inevitably there were darker jealousies working against Gagarin's peace of mind, and not just about the material fringe benefits that he was assumed to be enjoying. Sergei Belotserkovsky observes, 'Even Korolev couldn't have anticipated the avalanche of problems that would hit Gagarin when he had to represent his country abroad. He made many enemies because he behaved with more charm, and could talk more wisely and honestly, than the official Soviet heads of foreign delegations. Superiors never forgive you for something like that.'

Russayev worked hard to protect Gagarin from such dangers. 'He always said that politics seemed hard and intricate. I told him, "Politics is a dirty business. You should stay out of it. You've got your country, your family. Enjoy what you've got, and don't get involved in the politics."'

Russayev remained with Gagarin until 1964, when Khrushchev's administration was toppled by Leonid Brezhnev. After that, the KGB's relationship with all the cosmonauts would become very different. In March 1967 Gagarin would turn to Russayev one last time for some much-needed political guidance. By then it would be too late in the day for both of them.

10

BACK TO WORK

To their mutual surprise, Russian and American space medical experts had discovered by 1963 that the hardships of flight into space amounted to no more than a collection of minor irritations: nausea, vertigo, a heavy feeling in the head, a dryness in the throat. All these symptoms were uncomfortable, but any conventionally fit human could survive a trip into orbit. From now on the emphasis was on training not just the right bodies, but the right minds and intellects for working aboard an ever more complex succession of spacecraft. Korolev, Kamanin and other senior figures in the Soviet manned rocket programme re-examined the personal files of sixteen promising cosmonaut candidates previously rejected by the medical boards in 1959. They decided to give them another chance, because their engineering and academic skills were now regarded as more important than extreme physical fitness. By May 1964 this fresh cosmonaut squad had been supplemented with ten non-pilot technical specialists from within the space community itself. They were, in fact, talented engineers from within Korolev's OKB-1 design bureau. Meanwhile most of Gagarin's old friends and colleagues from the original 1959 group of twenty, including Gherman Titov, Alexei Leonov, Vladimir Komarov and Andrian Nikolayev, were studying very hard to maintain their superiority over the twenty-six newcomers.[1]

On December 21, 1963 Colonel Gagarin was appointed Deputy Director of the Cosmonaut Training Centre, reporting directly to Nikolai Kamanin. To a large extent this new job was a means of promoting him without putting him in harm's way. During the last three years he had fallen behind in every aspect of space training, and he had not been allowed to fly jet fighters because of the risks involved. Unlike a normal combat pilot, he was not expendable, but had to be preserved in one piece as a diplomatic and social symbol; even if he had been able to return to space-flight status at a moment's notice, the qualities that had made him so ideal for Vostok were no longer so important. It was not enough just to be a fit young pilot with the right attitude and background. If Yuri wanted to board a spacecraft again he would have to study orbital mechanics, flight systems, computer control and space navigation, then convince his superiors to put him back on the 'active' flight list. Korolev certainly wanted him back in the fold, but he had long ago warned his favourite 'little eagle' that he would have to get back into academic training as soon as possible.

As far back as June 1962 the Chief Designer had lost patience with the endless foreign trips and had complained, 'We're losing Gagarin and Titov as far as space is concerned.' He criticized Kamanin for failing to look after them properly. As so often, Kamanin deflected his own faults onto others, noting in his diary, 'Looking into his complaints, one can see Korolev's bitterness at having to keep his name under cover.'[2]

Korolev was already thinking beyond Voskhod, with plans for an ambitious new spacecraft capable of extraordinary feats: changing its orbit on command, adjusting its pitch and yaw attitude with millimetre accuracy and, most startling of all, making a rendezvous with another craft and docking with it to form a larger aggregate assembly. The new spacecraft was to be called 'Soyuz', meaning 'Union'. Of course this was a direct response to America's Apollo ship. In fact, Soyuz's general layout, with a rear equipment section, a re-entry capsule in the middle, and a dropaway docking compartment at the front, seemed suspiciously similar to an early proposal for Apollo drawn up by the General Electric Company in a failed bid to win a NASA contract.

Soyuz was a key element in future lunar plans, but it would not be ready for another two years at least. In the meantime, Konstantin Feoktistov, one of Korolev's most trusted engineers and a close colleague of Oleg Ivanovsky, was developing Voskhod as fast as was humanly possible, so that Korolev could fulfil his private 'deal' with Khrushchev. Feoktistov was also training to fly in Voskhod as the first specialist engineer-cosmonaut, along with the nine other engineers from OKB-1 who had passed the (by now much simpler) medical qualifications. Either this was Feoktistov's way of showing faith in his own work or it was Korolev's gesture of thanks for developing Voskhod so quickly. Feoktistov was the only person within OKB-1 who never gave ground to the Chief Designer on technical matters. In their stubborn fearlessness, the two men were remarkably alike – the world having already done its worst to them. While Korolev came come close to death in a Siberian prison camp, Feoktistov fought for the Red Army and was captured by the Nazis. After a brutal interrogation, they lined him up against a ditch and opened fire. He fell onto a pile of dead bodies and hid under them until nightfall, until he could limp away. The Voskhod capsule cannot have held many terrors for him.[3]

As Feoktistov and other highly skilled and experienced men like him began to rise in the cosmonaut hierarchy, so the chances were lessened for Gagarin to catch up with his studies and earn another flight into space.

In between his foreign trips, Gagarin had attended cosmonaut lectures as often as possible, but on many occasions he took his seat in the classroom only to be called away at short notice for some diplomatic function or other. When Khrushchev's administration ran into trouble from 1963, Gagarin managed to extend his academic work because he was not required to be quite so much in the limelight. In March 1964 he came to the Zhukovsky Academy in Moscow, a renowned school covering all aspects of aviation and aerodynamics, housed in the elegant Petrovsky Palace on Leningradsky Prospekt. Catherine the Great built Petrovsky Palace as a rest-stop for royal travellers, and Napoleon sheltered within its crenellated walls during the fire of Moscow in 1812. Now it was a necessary stop for cosmonauts on their way up into space. A special course had been established for the new science of space flight: the 'Pilot-Engineer-Cosmonaut Diploma'. Candidates would have to study all aspects of space and contribute a thesis in a chosen field of specialization, which they would then defend before their tutors in written and oral sessions at the end of the course. The cosmonauts were becoming more like their American counterparts, who also undertook diploma work to qualify themselves for space flight. (In a notable thesis, NASA recruit Buzz Aldrin outlined the mathematics for orbital rendezvous, thus earning himself a secure ranking in the astronaut corps.)

With every intention of making his mark at Zhukovsky, Gagarin selected for his thesis nothing less than the holy grail of manned space flight: a practical design for a reusable winged space plane. Alexei Leonov, who studied with him for several months, recalls, 'He was very strict with himself. I was always amazed how conscientious he was about his studies, how thoroughly and painstakingly he prepared his work, and how hard he tried to keep up with the others. Somebody who was full of airs and graces wouldn't have put himself through all that.'

Obviously a spaceship with wings could come home in an orderly fashion, landing at an airbase instead of falling down into a ploughed field or splashing into the sea. The wings would slow and control the final descent, so that the ship could touch down softly on wheels. Unlike the clumsy space capsules, a winged craft could be refurbished for another flight. The difficulty was to balance the aerodynamic usefulness of wings with the need for bulky re-entry heat-shielding. NASA had already started work on so-called 'lifting bodies', experimental craft that were neither capsules nor aircraft but something in between. They were dropped at great altitude from beneath the wings of B-52 bombers, and most of them landed successfully. However, it was impossible to send them into space because the addition of rocket engines and fuel tanks would have made them too heavy; and there was the intractable problem of the heat-shielding. At that time no sufficiently strong and lightweight material seemed capable of protecting the lifting body's stubby wings against melting away during re-entry. The 'ablative' heat-shielding of conventional capsules was thick and heavy, and it burned away irretrievably, leaving terrible scars on the capsule's flanks. The bulky resins and fibres used for the shields were completely unsuitable for wings.

In all, the spaceplane presented the most complex technical challenge. Even the current NASA space shuttle is a flawed design, consisting as it does of heavy components and throwaway tanks, with clumsy ceramic tiles to protect it against the heat. The search is still on for a genuinely efficient design. For Gagarin to research into these issues in the mid-1960s was proof of his great seriousness in attempting to re-qualify for space flight. Today few people remember his engineering skills; only his simple farmboy's smile. His painstaking and disciplined diploma work at the Zhukovsky Academy has been entirely forgotten except by his closest colleagues – in particular Sergei Belotserkovsky, the Deputy Director at Zhukovsky and the man who, more than any other, was responsible for the cosmonauts' academic skills in space flight and orbital dynamics (while Kamanin and the specialists at Star City taught them how to operate the hardware).

One of Gagarin's most significant achievements was to under-stand that, for safety reasons, his spaceplane had to be capable of an unpowered landing. Some of his tutors insisted that this was not technically possible. Gagarin argued that the spaceplane was useless if it could not make a 'deadstick' descent. After all, how could the crew get back if their engines failed? Just as for the Vostok capsules, a small braking motor should be enough to nudge the spaceplane out of orbit, he insisted; after that, it should be capable of reaching the ground without engines. His first solution was to bring the plane down by parachute, but of course that idea missed the point. Eventually he decided that it should glide to its landing. NASA's modern shuttles do precisely that, with no use of engines during their final approaches.

In some crucial aspects, Gagarin's thinking on the spaceplane concept outstripped that of his tutors; but on matters of strict aerodynamic science they pushed him hard. How would his spaceplane react to tail winds, head winds and cross winds? What about sudden, brief gusts? Had he calculated the drastic change of airflow as the ship neared the ground? Time and again Gagarin ran complicated mathematical simulations on primitive analogue computers (alongside his friend and collaborator Andrian Nikolayev), attempting to mate the airflow numbers to his ideal design. He revelled in the cut-and-thrust arguments with his tutors and collaborators, constantly striving to improve his ideas.[4]

Gagarin's spaceplane work was considered top-secret – in fact, all of the diploma work conducted at Zhukovsky was highly classified. Just as for Korolev, Belotserkovsky's identity was a secret and he was never acknowledged in public. He was not even allowed to take snapshots of his favourite pupils, in case Western spies should identify them. Even so, he used a hidden camera to obtain his keepsakes. 'We hid all the films in a safe, and developed them a long time later. I don't think we did anything wrong. Thanks to our unofficial action, we have an important and historical collection of photographs today.'

Gagarin became so absorbed in his work that he spent long periods staying at the Academy's hostel, instead of with Valya and the children. But the pursuit of academic excellence was not the only thing keeping him away from home.

Gagarin also spent a good deal of time at the Yusnost Hotel, just outside the western perimeter of Moscow's Garden Ring road. The Yusnost was associated with the official communist youth movement, 'Komsomol'. Gagarin was always welcome there, and room 709 on the seventh floor was kept in reserve for him, funded on a semi-permanent retainer by the Komsomol. On frequent occasions, banquets and receptions would be held at the Yusnost for Komsomol delegations visiting Moscow from other republics in the Union, and Gagarin would be expected to attend and give rousing speeches. Often the parties would drag on until the early hours of the morning, at which point it seemed best for Gagarin to sleep at the Yusnost, instead of struggling home through the freezing Moscow nights. In the relative informality and privacy of the hotel, he could relax and entertain his friends. An excellent billiards player, he seldom lost, except on one notorious occasion when he surrendered a game to 'Nona', a young chess champion noted for her attractive appearance. Gagarin's male companions could not understand how he could bear the humiliation of losing to a girl, but he had another game in mind.

Gagarin was a fit and handsome young man, who also happened to be the most famous and desired star in the world, with the possible exception of the 'fab four' young lads of the Beatles pop group; but it would be a mistake to think of him as a heartless womanizer. He was neither more nor less a sexual adventurer than any other superstar might have been in his circumstances. By all accounts, he loved Valya and was utterly devoted to his two little children. But Valya was not content for her husband to take his wedding vows casually. A single act of adultery was sufficient to upset her, let alone the 'several' that must certainly have occurred throughout the couple's married life.

On one occasion Valya decided to visit her errant partner at the Yusnost Hotel. Gagarin's favourite barber at the Yusnost, Igor Khoklov, blames lax security at the front desk for what happened next. 'It was a different era in those days. Any woman would have jumped on Gagarin, walked with him, slept with him, even. He had a few opportunities at the Yusnost. After a party when he was tipsy, a leading sportswoman, a ski champion, took an interest in him. He didn't seduce her, she seduced him, but around six o'clock his wife arrived. Perhaps she'd had some kind of premonition? I would say the military police [at the front desk] were at fault, because they could easily have telephoned up to Yura's room to warn him, but they didn't. And she created an uproar, she really tore Gagarin apart. The other girl just picked up her clothes and ran. I'd say that sportswoman really cost Gagarin dear.'

For Valya, this must have seemed all too familiar – the bang on the door, the arguments, the wounds to her husband's face (this time caused by her own nails, rather than by a jump from a balcony). Next morning Khoklov had to disguise the scratches on Gagarin's cheeks prior to a meeting with Khrushchev. Khoklov recalls, 'I did his make-up. I was licking his wounds, so to speak. He was the kind of a man who had a taste for women, but I wouldn't say he was throwing himself around left and right. Valentina loved her husband very much, but because of all the things that had happened, she was very jealous. Plus the fact that she caught him.'

Gherman Titov says, 'It was very difficult for Yura's wife to get used to the fact that he didn't really belong to her any more.'

One of the restraining influences on Gagarin's behaviour after Foros was the bodyguard assigned to him, on Khrushchev's orders – a tall, sour-faced man nicknamed the 'polecat', who immediately dampened the mood of any room he walked into, although he was apparently very nice, once one got to know him. By 1962 Gagarin had persuaded Khrushchev to withdraw the guard, but he had merely replaced him with three rather poorly disguised undercover followers. Khoklov says, 'If Gagarin liked a woman he couldn't go with her. You can't put the bodyguards in the same bed. If you want a drink, it's just the same. You have to buy a drink for the bodyguards.'

The other major factor was sheer pressure of work, not only at Zhukovsky but also at Star City, and during official ceremonial and propaganda functions. These were reduced in the Brezhnev era but by no means eliminated. Khoklov recalls Gagarin arriving for a haircut one day and commenting about a drunken tramp lying on the street outside, who had soiled his trousers. Exasperated by his relentless schedule, Gagarin joked bitterly, 'That's an intelligent fellow over there. He finds an opportunity to rest and do all the other stuff at the same time.'

Igor Khoklov was often despatched to the Kremlin to cut Khrushchev's hair. The wily old barber retains an unflattering memory of the endless KGB staffers and secret policemen he encountered in those days. 'I was in the room with Khrushchev and his special bodyguard, who kept his hand in his pocket where he hid his gun. I was thinking, "Who's faster, you with your gun or me with my razor?" A senior manager then came in, a good bloke, Armenian I think. He saw the guard and said, "We trust the barber. With Igor here, a guard isn't needed." So the guard had to go and wait outside the door.'

And it wasn't Igor the barber, but Khrushchev's closest political colleagues who eventually cut his throat.

Khrushchev demonstrated that the Soviet Union was a modern technological nation, a powerful player on the world stage, with its missiles, space rockets, satellites, computers, jet planes, aircraft carriers and nuclear weapons. His weak spot – as for so many Russian leaders before or since – was providing the people back home with the most basic necessity of life: food. In the autumn of 1963 he was forced into an embarrassing series of emergency measures, buying wheat from America to make up for the poor harvests yielded by his over-ambitious and under-planned 'Virgin Lands' grain-planting programme.

In a similar manner to President Kennedy in the US, Khrushchev looked to the glamour of space to divert attention from his failures. On October 12, 1964, Korolev made good on his promise, successfully launching Voskhod I with three crewmen aboard. Vladimir Komarov, Konstantin Feoktistov and Boris Yegorov were denied ejection seats in the cramped cabin, and there was not even room for them to wear spacesuits. Instead, they had to make do with simple cotton coveralls. However, the Voskhod incorporated some useful improvements on the old Vostok design. There was a back-up retro-rocket pod at the front of the craft, just in case the primary unit failed, and the re-entry capsule had a slightly flattened underside with a cluster of small rockets to soften its impact with the ground, thereby allowing the crew to stay aboard all the way to touchdown. Soviet spokesmen proudly heralded this new 'soft-landing' concept, obviously forgetting all the stories they had told about Gagarin's very different homecoming technique in 1961.

Voskhod I flew too late to benefit Khrushchev. The capsule came home on October 13, and on the very next day the First Secretary was deposed. In fact, he was called away from his retreat at Foros even as the flight was in progress. In Moscow a special meeting of the Politburo informed him, much to his surprise, that he had just resigned due to age and increasing ill-health. Khrushchev's deputy Leonid Brezhnev took advantage of the unresolved grain crisis to take over as First Secretary. Khrushchev's faithful aide Fyodor Burlatsky says today, 'Never mind the coup which happened recently during Gorbachev's time. This was a real coup, prepared by the KGB and Brezhnev against Khrushchev and against anti-Stalinism.'

Almost immediately Gagarin's status was affected. His foreign trips were curtailed and his lines of communication with the Kremlin severed. Brezhnev did not care to be reminded of his predecessor's space triumphs. Burlatsky, who knew Gagarin very well, noticed an immediate change in the young cosmonaut's mood. 'I'm sure that he became unhappy. It wasn't because he disliked Brezhnev. No, quite the opposite, it was because Brezhnev regarded him as a representative around the world for Khrushchev. Immediately Gagarin lost his status, his position. I had the feeling he didn't know what to do with himself. Politically he represented the hand of peace extended from the Soviet Union to the West, but Brezhnev started up the arms race again, and he didn't need people like Gagarin.' Burlatsky stresses the perpetual truism of political life in Russia. 'It's not so important who's who, but who belongs to whom. Gagarin belonged to Khrushchev, and that was enough to finish his career in Brezhnev's time.'

Burlatsky is not alone in his opinion that the new hard-line regime 'affected Gagarin's life in such a way that he lost everything, and he had to try and find himself again in some kind of new experience. Drink, perhaps. He was devastated. One day he was a representative of his country, and the next, a simple pilot without any position. Somebody once wrote, "The greatest unhappiness is to have known happiness before." Brezhnev and his Politburo friends took that happiness away, and they were guilty of everything that happened to Gagarin afterwards.'

Gagarin's personal driver Fyodor Dyemchuk sums up the political fall-out for the First Cosmonaut in the most straight-forward way. He remembers that in the Khrushchev days Gagarin's frequent trips to the Kremlin were happy affairs, often accompanied by laughter and drink. In Brezhnev's time the trips became far less frequent, 'and Gagarin would come out looking sad and sit quietly in the car. I wouldn't ask him what was wrong. I didn't need to. I could see he was busy with his thoughts.'

Gagarin's greatest initial shock was that he was no longer able to work behind the scenes on behalf of the many people who came to him with pleas for help. He was hardly a saint, but he was without doubt an essentially good-natured man, a product of his decent upbringing in Gzhatsk and Klushino. The virtues of social and personal responsibility that he had learned during the war years stayed with him throughout his life. Among those former colleagues who hint today at Gagarin's moments of caprice, his occasional misbehaviour and thoughtlessness, none of them denies his warm and generous behaviour towards friends and strangers alike when they were in trouble. In fact, by 1964 all the cosmonauts who had completed missions were well known, and their influence stretched a long way in higher quarters.

At Star City a special Correspondence Department was established ten days after Gagarin's flight to deal with the immense quantities of mail coming in from all over the Union, and many foreign countries besides. Over time the department was expanded to deal with other cosmonauts' correspondence, with seven secretaries on permanent duty (at least two of whom were answerable to the KGB). Sergei Yegupov headed the operation with two principal aims in mind: first, to help Gagarin with his workload; second, to keep an eye on any sensitive matters that might crop up. His duties are more politically relaxed now, and the world's fascination with cosmonauts has faded, but he still runs the department. 'Most often the letters were addressed to "Gagarin, Moscow", or "Gagarin, the Kremlin". In the end it was decided to give him a special post code, "Moscow 705". Over the years, I think we must have received at least a million letters.'

Most of the letters – but by no means all of them – expressed joy, wonder, admiration and pride at Gagarin's achievement. Yegupov does not like to acknowledge that some of these letters must have been 'difficult' in some way. 'We still keep the entire archive at Star City, and everybody can get access to them. You will see, there are no bad letters anywhere in the correspondence.' It seems reasonable to assume that some of them must have been weeded out, but a great many distressing pleas for help still remain.

'About ten or fifteen per cent of the correspondence contained various requests from ordinary citizens asking for better housing, the installation of water supply points, an increase in pension payments, and applications for kindergartens – incidentally, that was a pretty complicated issue in those days.'

The most sensitive letters came from prisoners asking for their cases to be reviewed. Fyodor Dyemchuk remembers a particular incident involving a young man harshly scntcnccd for a first offence. 'Gagarin said, "What shall I do? I have to help, because if we save this boy it'll be so much easier and simpler. If he goes to prison he'll simply be lost for ever into the crime system, and he'll never make it to become a man." He went everywhere and visited everybody's offices. He really pushed, and I think he got some positive results.'

One can only imagine Gagarin's anxiety when faced with requests like these:

> Esteemed Yuri Alexeyevich Gagarin,
> A First Class Navigator who served with the Air Force for nineteen years requests you to receive him. The life of my son depends on this . . .

> Yuri Alexeyevich, Hero of the Soviet Union,
> My daughter has been refused entry to the university because of my Jewish background. Please can you . . .

> Dear Comrade Gagarin,
> Citizen Danilchenko asks you to consider helping to obtain a reservation for his daughter at an invalid's residential home, since she is mentally unwell . . .

Yegupov pulls out other typical examples from the archive. 'Here's a letter asking for better housing for the Kurdyumov family, nine people living in a single room, sixteen metres square, in an old house with damp running down the walls. Here's another request for better housing from citizen Morazova, and meanwhile she has a child with an inborn heart deficiency; or this one, where prisoner Yakutin says he was wrongly convicted, and requests the verdict to be cancelled.' Somehow Gagarin managed to answer almost all the letters, showing great concern, and even folding a particularly heart-breaking example into his wallet as a spur to his conscience. He was, however, entirely unmoved by some of the pleas he received:

> Dear Cosmonaut Yuri Gagarin,
> Please allow me to congratulate you on your great feat. Please allow me also to ask a great favour on behalf of the distilling firm of Richter & Co., and that is to be so kind as to allow us to use your well-known and respected name for a new product, 'Astronaut Gagarin Vodka'.

'Why did you have to show me this?' he complained to Yegupov. 'I've wasted three minutes on reading it.' Then he spent a good half-hour phrasing a careful and warm reply to a 15-year-old boy from Canada who had politely written in for career advice.

As often as not, people would actually accost the First Cosmonaut in person with their pleas for help. Alexei Leonov says that whenever Gagarin visited his family in Gzhatsk, he would find various local dignitaries waiting for him in his parents' house with requests for political favours. 'He did a lot for his old neighbours and the people of Smolensk. Gzhatsk was originally an old-fashioned merchants' town, but after 1961 it started to bloom, and it turned into a modern, highly developed city.' Gagarin's name and reputation stimulated a drastic change of fortune for the entire region.

But there was one notable occasion when Gagarin refused point-blank to help. A mother wrote to him saying that her son was in trouble for cutting down a fir tree in a forbidden area at Christmas time. Gagarin looked into the business, found out that it had probably been more than one tree and that the young man was selling them off for profit. He recommended the man be sacked from his job. According to his driver, Gagarin became pretty angry and said, 'What happens if everyone goes and cuts down "just one" fir tree? Where are we going to live then? Any day now, we won't have anything left.'

Leonov puts this (and other similar incidents) down to Gagarin's perceptions of the earth from space. 'After his flight he was always saying how special the world is, and how we had to be very careful not to break it.' This is a common enough truism by modern standards, taught to all of us in school, but what must it have been like for the very first man in space to discover it for himself? In April 1961 Gagarin was the only human being among three billion who had actually seen the world as a tiny blue ball drifting through the infinite cosmic darkness.

So the tree-cutter lost his job at Gagarin's specific request, but more often he was inclined to help his petitioners by appealing to higher authorities. 'You could hardly find a single man who wouldn't assist him if he asked for it. Who could refuse him?' says Yegupov.

Leonid Ilyich Brezhnev could.

In his first months in office, Brezhnev was preoccupied with achieving dominance over his co-conspirator Alexei Kosygin. Brezhnev's attitude towards Korolev was similar to Khruschev's: an insistence on orbital 'firsts', accompanied by a hazy lack of interest in the exact technical details. However, the scheduled mission of Voskhod II did interest Brezhnev, because it promised a major new triumph: the first spacewalk, enabled by a flexible airlock attached to the re-entry ball's flank. Korolev was just as keen to try out this new concept. In 1962 he prepared Leonov, who was one of the prime candidates for the first walk, with a suitable pep-talk. 'He told me that any sailor has to learn to swim, and each cosmonaut has to know how to swim and do construction work outside his vehicle.'

On February 23, 1965, Korolev launched an unmanned test vehicle with the new airlock attached. The mission ended badly when the capsule broke up during re-entry, as a result of poor command signalling from the ground. A few days later, an air-drop of the capsule from a plane also failed because the parachute did not open. Oleg Ivanovsky remembers Korolev saying in disgust, 'I'm sick of flying under rags.' He hated parachutes and always wished that he could design a rigid rotor system, or some other aerodynamic device to replace them. Perhaps it was a mercy that he never lived to see a much more terrible parachute failure in April 1967: a failure that might easily have claimed the life of Yuri Gagarin . . .

The Voskhod II mission took off on March 18, 1965 with wonderful timing, ahead of NASA's first Gemini mission by just six days. This time there were only two crewmen in the cabin, to make room for their bulky spacesuits. Pavel Belyayev remained inside, while his co-pilot Alexei Leonov squeezed into the flexible airlock and pushed himself out of the capsule. For ten minutes he enjoyed the exhilarating sensation of spacewalking, and then began to pull himself back into the ship – only to discover that his suit, at full pressure, had ballooned outwards, so that he could no longer fit into the airlock. Extremely exhausted by his efforts, Leonov had to let some of the air out of his suit to collapse it, so that he could squeeze back aboard.

Then, prior to coming home the next day, Belyayev saw that the ship's attitude was incorrect for the braking burn, and he shut down the automatic guidance systems before they could make matters any worse. With help from Korolev and ground control, he and Leonov had to ignite the braking motors manually on the next orbit, displacing their eventual landing site by 2,000 kilometres. The capsule descended onto snow-covered wilderness near Perm, alongside the very northernmost reaches of the Volga. It smashed into a dense cluster of fir trees and was wedged several metres off the ground between two sturdy trunks. Meanwhile the recovery teams were 2,000 kilometres away, in the zone where they had expected the capsule to come down. The cosmonauts had to spend a restless, frozen night waiting to be picked up. They pushed open the capsule's hatch but dared not climb down from their precarious perch, because a pack of wolves was howling somewhere very nearby in the darkness.[5]

These difficulties were not mentioned in the Soviet press reports, and the mission was a great propaganda triumph all around the world, particularly after Yuri Mazzhorin had responded to a stern phone call from Alexei Kosygin's office. 'They said that not a single word about the landing in Perm should appear in the media. I had no idea what that region looked like, but I had to go to all the television stations and make sure you couldn't identify Perm in any of their news footage.' Subterfuge aside, the fact remains that Alexei Leonov walked in space well ahead of his American rivals. The London *Evening Standard* ran an article about the US astronauts Young and Grissom, gearing up for the first Gemini shot, with the headline 'FOLLOW THAT CAB!', while *The Times* described Leonov's adventure as 'a fantastic moment in history'. Once again NASA had been trumped. Leonov's spacewalking rival, Ed White, did not get his own chance to catch up until the second Gemini mission on June 3, nearly three months after Voskhod II's flight.

An accomplished artist, Leonov set about designing a commemorative postage stamp showing his spacewalk, and spent hours happily chatting with Gagarin about the differences each man had observed in the curvature of the earth. 'My view was much steeper, much rounder than Gagarin had reported, and it worried me, but then we realized that Voskhod's maximum orbital altitude was 500 kilometres, and Vostok's 250 kilometres, so I was much higher. You see, everything has a sensible physical explanation.' It was the infernal rules of secrecy that made no sense. Leonov's innocent stamp designs had to be vetted by the KGB propaganda experts. 'Everything was so secret – they were all civil servants, you see. I drew a completely different spacecraft that wasn't anything like [the one that had really flown] and then they were satisfied.'

In the wake of this more-or-less successful mission, Voskhod III was scheduled to coincide with the Party Congress of March 1966. Cosmonauts Georgi Shonin and Boris Volynov began training for an ambitious rendezvous with an unmanned target vehicle. Even the journalist Yaroslav Golovanov was recruited for forthcoming Voskhod missions, along with two other writers, after Korolev had expressed frustration at the cosmonauts' rather prosaic descriptions of space flight.

On January 14, 1966 Korolev was in the Kremlin Hospital for a supposedly routine intestinal operation. Weakened by years of ill-health and overwork, and by the great damage inflicted during his imprisonment in a Siberian labour camp from 1938 to 1940, his body was far more fragile than the doctors had suspected. Internal bleeding proved difficult to control, and two huge tumours had developed in his abdomen. After a lengthy and fraught operation, Korolev's heart gave out and he died.

Gagarin was furious that the privileged and well-rewarded doctors had not been able to save his mentor and friend. Sergei Belotserkovsky remembers him raging, 'How can they treat someone so respected in such a mediocre and irresponsible way!' In fact, Yuri had always said that he did not trust the Kremlin's special hospital for the élite. In June 1964 Valentina Tereshkova had been assigned a room there so that she could give birth to her baby (seven months after her marriage to Andrian Nikolayev). Gagarin's barber friend Igor Khoklov remembers his carefully reasoned distrust. 'He said, "None of those old bosses in the Politburo are capable of fathering babies any more. The Kremlin hospital delivers maybe one or two a month. We must send Valentina to an ordinary people's hospital where they deliver babies on a conveyor belt, and have the experience to know what they're doing." He had a very good relationship with Tereshkova, by the way.' Gagarin's wishes were granted and the Kremlin hospital lost its star female guest. Now, in the bleak and bitter-cold January of 1966, Gagarin was deeply distressed by Korolev's death, and angry.

Throughout all his years working to give the Soviets a lead in space, Korolev had never discussed his arrest, torture, beatings and imprisonment under the old Stalinist regime. People thought of him as a burly man, built like a bear, but the truth was that his body was made rigid by countless ancient injuries. He could not turn his neck, but had to swivel his upper torso to look people in the eye; nor could he open his jaws wide enough to laugh out loud.

Two days before he was scheduled for surgery he was resting at his home in the Ostankino district of Moscow. Gagarin and Leonov came to visit him with several other colleagues, and at the end of the evening, just as most of the visitors were putting on their greatcoats to leave, Korolev said to his two favourite cosmonauts, 'Don't go just yet. I want to talk.' So his wife Nina fetched some more food and drink, and for four hours, well into the early hours of the morning, Korolev told the story of his early life – a story that Leonov has never forgotten. 'He told us how he was arrested, taken away and beaten. When he asked for a glass of water, they smashed him in the face with the water jug . . . They demanded a list of so-called traitors and saboteurs [in the early rocket programme] and he could only reply that he had no such list.' Korolev described how he saw, through puffy eyes, that his captors had pushed a piece of paper between his bruised fingers for him to sign; how they beat him again, and sentenced him to ten years' hard labour in Siberia. 'Yuri and I were both struck by the unexpected parts of his story,' says Leonov.

From a living death in Siberia, Korolev was recalled to Moscow when an old ally of his, the renowned aircraft designer Andrei Tupolev, requested him for war work. He would be assigned to a less harsh special prison facility for engineers, which included design offices and better living conditions.[6] In fact, Tupolev was himself a prisoner. But no special arrangements were made to transport Korolev to Moscow, and he had to improvise. Cold beyond endurance and hallucinating with hunger, he found a hot loaf of bread on the ground one day, apparently dropped from a passing truck. 'It seemed like a miracle,' he told Gagarin and Leonov. He worked as a labourer and shoe repairer to earn his passage back to Moscow by boat and rail. His teeth were now loose and bleeding, because he had not eaten fresh fruit or vegetables in a year. Trudging along a dirt track one day, he collapsed. An old man rubbed herbs on his gums and propped his body to face the feeble sun, but he collapsed again. As Leonov vividly recalls, 'He told us he could see something fluttering. It was a butterfly, something to remind him of life.'

It seems that after so many years' silence the ailing Chief Designer now wanted to unburden himself to his two favourite young friends. The two cosmonauts were deeply affected by what they heard. Leonov says, 'This was the first time that Korolev had ever talked about his imprisonment in the Gulag, since these stories are usually kept secret . . . We began to realize there was something wrong with our country . . . On our way home, Yuri couldn't stop questioning: how could it be that such unique people like Korolev had been subjected to repression? It was so obvious that Korolev was a national treasure.'

After the funeral Gagarin insisted on spending the night at Korolev's house. According to Yaroslav Golovanov, 'Gagarin said, "I won't feel right until I've taken Korolev's ashes to the moon." At the crematorium he asked cosmonaut Vladimir Komarov to scatter some of Korolev's ashes on the next space flight, during the descent, although, according to Orthodox custom, you must not divide a person's ashes.' It is not clear whether any ashes were actually taken into space, but Golovanov insists that several handfuls went missing from the crematorium. 'Komarov did scatter some of them after Korolev's death. Gagarin and Leonov also had some ashes.'

Korolev's death marked a turning point for Gagarin. He became totally recommitted to flying in space, and even to flying to the moon. His self-discipline returned, and he worked on his diploma with passionate energy. He impressed Kamanin, who allowed him to train as back-up for the first Soyuz mission. All being well, the back-up position would automatically entitle Gagarin to the second Soyuz flight. But this put him in direct conflict with another cosmonaut, who believed that the assignment belonged to him, not to Gagarin. The noted space historian James Oberg says, 'This tension between Gagarin and one or two of the cosmonauts isn't written about very much, perhaps because people don't like to talk about it. Basically Gagarin pulled rank.'[7]

11

FALLING TO EARTH

Most of the cosmonauts got on well with Gagarin, appreciated his humour and generosity of spirit, enjoyed drinking and partying with him and deferred to him as the undisputed leader of their cause. Many were anxious to see him back in space, but one cosmonaut in particular felt differently.

Georgi Timofeyevich Beregovoi, born in April 1921, was one of the oldest cosmonauts, recruited in 1963 when the list of 'near-misses' among the original 1959 candidates was re-evaluated. Among all the cosmonauts, only he and Pavel Belyayev (Leonov's commander on the Voskhod II mission) could claim the greatest distinction for a pilot: experience in real aerial combat. Beregovoi flew 185 missions against the Germans during the war, and was awarded the much-prized title Hero of the Soviet Union. During the 1950s he served as a test pilot, so when he signed up for cosmonaut training he believed he was well qualified. In 1964, quite soon after his selection, he gained a back-up posting for the planned Voskhod III mission and trained with every confidence that he would fly whatever mission came next. Nikolai Kamanin felt at ease with a fellow war veteran. He sponsored Beregovoi's recruitment and gave him every chance to succeed.

After Korolev's death, his deputy Vasily Mishin took over the administration of OKB-1. He was good-natured and eager, but he lacked the political influence and raw cunning of his predecessor.[1] One way and another, the Voskhod III schedule slipped so badly that it had to be cancelled. Mishin decided to focus OKB-1's energies on Soyuz and on the further, very troubled development of the large moon booster, the N-1.

189

Beregovoi now expected his back-up status to be shuffled smoothly along to the next mission, the first manned test of the new Soyuz configuration. At this point Gagarin stepped in and claimed that posting for himself, making every possible use of his rank as Deputy Director of Cosmonaut Training to do so. Beregovoi made his annoyance crystal-clear to anyone who would listen, and eventually stormed into Gagarin's office at Star City for a direct confrontation. Gagarin's driver Fyodor Dyemchuk walked into the office at the wrong moment and overheard the row. 'The other man was the superior in years, but he hadn't flown into space yet. He made indecent remarks about Gagarin, and said he was too young to be a proper Hero of the Soviet Union and he'd become conceited. He called Gagarin an upstart, and Gagarin replied, "While I'm in charge, you'll never fly in space." They argued for quite a while.'

There seems to have been some fault on both sides. Beregovoi was not automatically entitled to a Soyuz mission, despite what he may have thought. His training on the very different Voskhod hardware was completely inappropriate for the new ship, and he had no right to take out his run of bad luck on Gagarin, just because the Voskhod series had been brought to a close before he could fly. Judging from the fact that he rose to become Head of Star City in 1972, Beregovoi was an ambitious man.

After this row Dyemchuk reported fearfully to his master that some other cosmonauts and senior staff in the Star City compound had asked to use Gagarin's official car, and Dyemchuk, as a humble driver, had not been able to refuse them. 'He hit the car with his fist and said, "We have one commander here, and he's the only person who can order the car!" Apart from him, no one could reserve it. He showed his emotions and a small dent was left in the bonnet.' Dyemchuk is convinced that this outburst was entirely out of character for Gagarin, the result of stress rather than vanity.

Yaroslav Golovanov says, 'Around this time you heard other cosmonauts saying, "So what, about Gagarin? He flew once around the earth, and all he had to do was watch over Vostok's automatic systems." Well, that's not right, because when he flew, the whole business [of space flight] was starting from scratch, and everything he did was extremely important and brave, because nobody knew what might happen. They didn't even know if a man could swallow properly in space, or endure the lack of gravity. It's completely wrong to reproach Gagarin.'

Almost certainly these critics, muttering under their breath, were not from the first group of twenty, but from the new intake training for Soyuz. Not only had Korolev's death deprived Gagarin of a dear friend and mentor, it had also taken away his most important political protector in the space community, just as Khrushchev's fall from power had left him defenceless against jealous generals in the Kremlin. The First Cosmonaut had to fight unusually hard to maintain his position in the cosmonaut hierarchy, and the strain was wearing away at his old easy-going nature. Meanwhile OKB-1 was becoming a weaker bureau under the leadership of Vasily Mishin, who could not defend himself as effectively as Korolev against interference from the Kremlin, or from the vicious competition of rival aerospace bureaux anxious to increase their own profile in space.

Suddenly NASA's gigantic moon programme came to a halt. On January 27, 1967 Gus Grissom and his crewmates, Ed White and Roger Chaffee, clambered into the first flight-ready Apollo, atop a half-sized version of the Saturn rocket. This was supposedly a routine check-out procedure, during which they would run a simulated countdown with all systems running and bring the ship to the very last second before take-off, without actually igniting Saturn's engines. Morale around the pad was poor, even before the test began, because the new capsule had not come up to expectations. The detail work on the electrical and communications systems was inadequate, prompting the astronauts to stick a mouldy lemon on top of the capsule's duplicate simulator to show their contempt for the overall design. When Chaffee climbed through the hatch of the flight vehicle to start the test, he complained that the interior smelled of sour milk. The consensus was that the balky environmental control hardware was generating fumes. Then the radio system glitched.

Furious, Grissom shouted, 'How the hell are we supposed to communicate with mission control from space when we can't even talk to them on the ground!' The mood around the Kennedy launch complex was distinctly strained as the technicians locked Apollo's heavy hatch into place, sealing the crew inside.

Five hours into the test, Grissom's garbled voice on the crackling radio link said, 'We've got a fire in the capsule.' A few seconds later, another voice (possibly White's) was more urgent. 'Hey, we're burning up in here!' There was a scream of pain, then just a hiss of static as the radio went dead. Suddenly the side of the capsule split open. There was a horrifying 'whoosh!' as the top of the launch tower was engulfed in thick, acrid smoke and flames. The pad crew, high atop the gantry, tried desperately to get the astronauts out, but the smoke was impenetrable and the heat quite overpowering. It took four minutes to open Apollo's hatch, by which time all three astronauts were dead.[2]

The tragedy was reminiscent of Valentin Bondarenko's death in the isolation chamber back in 1960, but NASA could not draw any lessons from that because of the obsessive secrecy that always surrounded the Soviet space effort.

NASA entered a two-year hiatus, a period of self-doubt, its technical and political reputation severely tarnished by the deaths. The Soviet cosmonauts grieved for their US counterparts, and were permitted to send official expressions of condolence to the dead men's families, even as Leonid Brezhnev and Vasily Mishin speculated about the window of opportunity that had arisen for the Soviet space effort to take advantage of NASA's enforced slow-down.

The Soviets' gigantic N-1 lunar superbooster was running badly behind schedule, and even the demoralized Americans knew that it did not present a serious threat. According to a National Intelligence Estimate document of March 2, 1967, their appraisal of the N-1 was that:

Several factors militate against the Soviets being able to compete with the Apollo timetable . . . Their lunar launch vehicle will probably not be ready for test until mid-1968, and even then we would expect to see a series of unmanned tests lasting about a year to qualify the system before a lunar landing might be attempted. In the meantime they still have to test rendezvous and docking techniques.[3]

The landing and return of a man on the moon seemed very far off yet for Mishin and his beleaguered team at OKB-1; but a much simpler circumlunar flight, Jules Verne-style, might be achievable without the need for the colossal and as yet unflown N-1 booster. Korolev's old rivals in the aerospace community, Glushko and Chelomei, were developing a rocket called Proton, which was larger and more powerful than the R-7 but not quite powerful enough to carry a lunar landing module as well as a crew return capsule. Mishin faced a difficult choice: if he opted for the circumlunar flight aboard Chelomei's Proton, he would have to sacrifice some development work on the N-1 and the bug-like landing craft; but if he could achieve a fairly basic 'once-around-the-moon' flight while NASA was still preoccupied with recovering from the Apollo fire, any subsequent walkabout on the lunar surface by American astronauts would come across as yet another second-best. With this tantalizing prize in mind, work on the new Soyuz capsule was accelerated, while the N-1 was allowed to fall further behind and Korolev's old enemies dug their claws deeper into OKB-1.

The cosmonaut team was affected by all these complications. Leonov began training a squad for a touchdown mission with the N-1, which would include the use of the tiny one-man lunar landing pod, while another team of astronauts was prepared for circumlunar flights aboard the Proton, using an elongated Soyuz variant known as a 'Zond'. Meanwhile, yet another group, including Gagarin, was training for the first basic earth-orbital test of the Soyuz mated to a standard R-7. In contrast to the overall NASA effort, with Apollo as its privileged centrepiece, the Soviet lunar programmes were divided, confused and contradictory, especially in the absence of Korolev's managerial discipline.

By the spring of 1967, development of the Soyuz was moving towards that crucial first flight. On April 22 the Soviet propaganda departments felt confident enough to let slip some rumours to the international press agency, UPI. 'The up-coming mission will include the most spectacular Soviet space venture in history – an attempted in-flight hook-up between two ships and a transfer of crews.' But some doubts seemed to be preying on Nikolai Kamanin's mind. His diary more or less implies political pressure to push the Soyuz launch schedule forward:

> We must be fully convinced that the flight will be a success. It will be more complicated than previous flights, and the preparation will have to be appropriately longer . . . We do not intend to rush our programme. Excessive haste leads to fatal accidents, as in the case of the three American astronauts last January.[4]

Kamanin's anxiety presaged disaster. Alexei Leonov says, 'The first manned test of the Soyuz was assigned to Vladimir Komarov, with Yuri Gagarin as the back-up, and another Soyuz spacecraft was being prepared for Yuri to fly at a later date. He trained very hard for two years, reporting the progress of his training in detail to the State Committee. Then Komarov flew for two days [specifically, 27 hours] and we had a big problem.'

Komarov's launch was supposed to be followed a day later by another Soyuz with three more crewmen aboard: Valery Bykovsky, Yevgeny Khrunov and Alexei Yeliseyev. The two Soyuz ships were supposed to dock, then Khrunov and Yeliseyev were scheduled to spacewalk into Komarov's capsule and sit in his spare seats, thus producing another world 'first' – going up in one ship and coming home in another. This was designed as a rehearsal for a future moon mission. The Soyuz did not yet include an airtight docking tunnel, so the only way of swapping a crewman between the capsule and a future lunar lander would be to spacewalk him from hatch to hatch.

It seems likely that the Brezhnev administration wanted the docking to take place on or around May Day. The year 1967 had a special significance in the communist calendar; it was the fiftieth Anniversary of the 1917 Revolution. The concept of

making a 'union' between two spaceships collaborating in orbit
was highly symbolic, especially for a ruling government obsessed
with symbols. In 1982 Victor Yevsikov, an engineer on the Soyuz
development team who helped design the heat-shield, admitted
from his new safe haven in Canada that heavy political pressure
was applied to Vasily Mishin and OKB-1 to get the two Soyuz
ships into orbit on time:

> Some launches were made almost exclusively for propaganda
> purposes. An example, timed to celebrate International
> Solidarity Day in 1967, was the ill-fated flight of Vladimir
> Komarov . . . The management of the OKB-1 Design Bureau
> knew that the Soyuz vehicle had not been completely
> debugged, and more time was needed to make it operational,
> but the Communist Party ordered the launch, despite the fact
> that four preliminary unmanned tests had revealed faults . . .
> The flight took place despite Vasily Mishin's refusal to sign
> the endorsement papers for the Soyuz re-entry vehicle, which
> he considered unready.[5]

As the deadline for the mission drew near, OKB-1 technicians
knew of 203 separate faults in the spacecraft that still required
attention. Yuri Gagarin was closely involved in this assessment.[6]
By March 9, 1967, he and his closest cosmonaut colleagues had
produced a formal ten-page document, with the help of the engineers, in which all the problems were outlined in detail. The trouble was, no one knew what to do with it. Within Soviet society, bad
news always reflected badly on the messenger. Quite apart from
Mishin, as many as fifty senior engineers knew about the report,
or had helped draft it, but none of them felt sufficiently confident
to go into the Kremlin and do what had to be done: request that
Leonid Brezhnev play down the symbolism of the pending launch,
so as to allow a decent delay for technical improvements.

The cosmonauts and space bureaucrats eventually adopted an
age-old technique. They recruited a non-partisan messenger from
outside the Soyuz programme to deliver the document for them:
Yuri Gagarin's KGB friend Venyamin Russayev.

'Komarov invited me and my wife to visit his family,' says Russayev. 'Afterwards, as he was seeing us off, he said straight out, "I'm not going to make it back from this flight." As I knew the state of affairs, I asked him, "If you're so convinced you're going to die, then why don't you refuse the mission?" He answered, "If I don't make this flight, they'll send the back-up pilot instead. That's Yura, and he'll die instead of me. We've got to take care of him." . . . Komarov said he knew what he was talking about, and he burst into such bitter tears. Of course he kept his emotions in check in front of his wife, but when we were alone for a moment he collapsed completely.'

Russayev could not be of much help on his own. Back at his desk in the Lubyanka the next morning, after a sleepless night, he decided to ask advice from one of his KGB seniors, Major-General Konstantin Makharov, a man he respected. 'Makharov's department dealt with space issues in relation to personnel. He used to work very closely with Korolev, but he was gone, and [his successor] Mishin wasn't the same kind of man. The guys in my department also contributed to this work, but Mishin was impossible to deal with, particularly when firm decisions were required. He always needed a lot of guidance . . . I went to Makharov's office and told him there was a serious problem with the rocket. He listened to me very carefully, and then he said, "I'm going to do something. In the meantime don't leave your desk today. Not even for one second." I kept my promise, and I'd only been back at my desk for a short while when he sent for me again. He gave me a letter, prepared by a team mobilized by Yuri Gagarin. Most of the cosmonauts took part in the research. Makharov told me to take the letter upstairs and see Ivan Fadyekin, Head of Department Three.'

This 'letter' consisted of a covering note, along with the ten-page document describing all 203 problems in the Soyuz hardware. Russayev insists, 'I didn't read it. I simply didn't have the time.' Just as likely, his instincts as a KGB man warned him that sneaking a look at the document might be very dangerous for him. As soon as he saw it, Fadyekin decided the same thing and dodged the responsibility straight away. 'I don't have the expertise for this.'

He redirected Russayev to a much more dangerous man in the Lubyanka: Georgi Tsinev.

Tsinev was a close personal friend of Leonid Brezhnev; in fact, he was related by marriage, and they had fought alongside each other in the war. If anyone could deliver an important message straight into the hands of the First Secretary, Tsinev could. Unfortunately for Russayev, things were not quite that simple. Tsinev was rising fast within the KGB, helped along by his powerful patron in the Kremlin. He was not going to allow any irritations to disturb that cosy relationship. 'While reading the letter, Tsinev looked at me, gauging my reactions to see if I'd read it or not,' Russayev explains. He had the inescapable feeling that Tsinev already knew the document thoroughly and was not remotely interested in its technical details. 'He was glaring at me very intently, watching me like a hawk, and suddenly he asked, "How would you like a promotion up to my department?" He even offered me a better office.'

Russayev was now in great danger. Tsinev was trying to buy him off with a promotion, at the same time as placing him in a department where he could be more closely monitored. If Russayev accepted the deal, he would lose any chance of helping Komarov and Gagarin's cause. On the other hand, if he rejected Tsinev's offer, the consequences did not bear thinking about. 'It was all part of the game, I suppose. I was very angry, but I couldn't let it show. I declined Tsinev's offer very carefully, explaining that I wasn't really qualified for the work in his department.'

Tsinev kept hold of the document and it was never seen again. Within weeks, Fadyekin was transferred to a junior consular office in Iran, merely for the crime of glancing through it. Makharov was fired immediately, without a pension, and Tsinev took over as chief of an entire counter-intelligence department. Russayev was stripped of any responsibility for space affairs, and transferred to an insignificant staff training department outside Moscow, well away from the Lubyanka. 'I kept my head down like a hermit for the next ten years,' he says.

* * *

Early on the morning of April 23, 1967, the Soyuz was propped up against the gantry at Baikonur, ready for launch, according to the original schedule. As Komarov made his final preparations before taking the lift up to his seat in the capsule, Gagarin seemed to have forgotten that the torture inflicted on back-up pilots in the old Vostok days no longer applied. Instead of forcing them into spacesuits and driving them all the way to the base of the pad to watch their more fortunate colleagues ascend to the top of the rocket, back-ups were now stood down from duty the night before a flight – only this time, Komarov was not necessarily the more fortunate man. The journalist Yaroslav Golovanov noticed Gagarin behaving very strangely. 'He demanded to be put into the protective spacesuit. It was already clear that Komarov was perfectly fit to fly, and there were only three or four hours remaining until lift-off time, but he suddenly burst out and started demanding this and that. It was a sudden caprice.' Golovanov did not realize that this was not random misbehaviour. Russayev and others insist that Gagarin was trying to elbow his way onto the flight in order to save Komarov from almost certain death.

The problem with Golovanov's version is that Komarov was not supposed to wear a spacesuit for this mission; therefore his back-up, Gagarin, would not have been assigned a suit, either. The front module of the Soyuz incorporated airtight hatches at each end, allowing the module to serve as an airlock. The spacewalkers from the second Soyuz needed their suits, but Komarov did not. In which case, why did Gagarin demand to be dressed in a suit? A more realistic explanation is that he wanted Komarov to wear the spacesuit in order to give him an extra safety margin. This is not as simple as it sounds; spacesuits are so much an integral part of a capsule's system that they cannot always be worn with the simplicity of an overcoat, but have to be plugged in to other life-support machinery. Another possibility is that Gagarin was trying to disrupt the preparations somehow, but without any clear plan of action. Whatever happened in the suiting-up room that morning, the archive pre-launch footage shows an unhappy Komarov, a downcast Gagarin and some very subdued technicians.

Komarov encountered problems almost as soon as he had

achieved orbit. One of the two solar-power vanes on the rear equipment module refused to deploy (yet another mechanical problem) and his guidance computers ran short of power. Launch of the second Soyuz, crewed by Yeliseyev, Khrunov and Bykovsky, was cancelled while ground controllers worked on Komarov's power deficit, although by some accounts Vasily Mishin held off cancelling the second launch for as long as he could. After eighteen orbits (twenty-six hours), Komarov's problems had not been resolved, and the mission directors decided to terminate the mission altogether during the next orbit. Komarov had great difficulty lining up his capsule for re-entry and complained, 'This devil ship! Nothing I lay my hands on works properly.'

Unlike the old Vostok ball, the Soyuz capsule had a distinctly flattened underside to give it some aerodynamic lift in the atmosphere, rather like an Apollo module. The drawback was that it had to be aimed much more accurately than Vostok. With his guidance systems almost entirely off-line, Komarov could not keep his ship at a stable angle, and when it began to spin, he fired his attitude control jets to try and bring it back in hand. Unfortunately the OKB-1 designers had put the thrusters too close alongside the star tracker navigation sensors, and the delicate lenses could no longer tell stars from random reflections. Passing over the night side of earth, and searching for a more obvious reference target for his blurred instruments, Komarov had to use the moon in a desperate attempt to work out his alignment.[7]

Rumours about the dialogue between Komarov and ground control have circulated for many years, based on reports from American National Security Agency (NSA) staff monitoring the radio signals from a USAF facility near Istanbul. In August 1972 a former NSA analyst, interviewed under the name 'Winslow Peck' (real name Perry Fellwock), gave a very moving account of the interception:

They knew they had problems for about two hours before Komarov died, and were fighting to correct them. We taped [the dialogue] and listened to it a couple of times afterwards.

Kosygin called Komarov personally. They had a video-phone conversation, and Kosygin was crying. He told him he was a hero . . . (The guy's wife got on too, and they talked for a while. He told her how to handle their affairs, and what to do with the kids. It was pretty awful. Towards the last few minutes, he was falling apart . . . The strange thing is, we were all pretty bummed-out by the whole thing. In a lot of ways, having the sort of job we did humanizes the Russians. You study them so much, and listen to them for so many hours, that pretty soon you come to know them better than your own people.[8]

As he began his descent into the atmosphere, Komarov knew he was in terrible trouble. The radio outposts in Turkey intercepted his cries of rage and frustration as he plunged to his death, cursing for ever the people who had put him inside a botched spaceship – although his 'final screams', mentioned later in Fellwock's account, may be an exaggeration.

Korolev's inadvertent prophesy about 'flying under rags' was fulfilled when the parachutes did not deploy properly. A small drogue canopy came out, but failed to pull the bigger canopy from its storage bay – yet another serious design flaw. A back-up parachute was released, only to become entangled with the first drogue. There was nothing to slow the capsule's fall, and Komarov slammed onto the steppe near Orenburg with all the force of an unrestrained 2.8-ton meteorite. The capsule was utterly flattened, and the buffer retro-rockets in its base blew up on impact, burning what little wreckage was left.

Recovery troops picked up handfuls of soil to try and dampen the flames. Their radio messages back to base were garbled and distressed: something about the cosmonaut 'requiring urgent medical attention'. It seems unlikely that any recognizable portion of Komarov's body would have survived intact, although Russayev says that a heel bone was found among the ashes.

This was the first Soviet fatality during an actual space flight, and it came as an immense shock; nor could the basic truth of the disaster be discreetly hidden from the outside world (although the Soviet authorities admitted only to an unfortunate parachute failure, and not to a series of design and preparation flaws dating from long before the ship took off). This time it was NASA's turn to send letters of condolence. Both sides in the superpower divide had learned that the space environment showed no concern for nationalities or flags, but treated all trespassers – Russian and American alike – to the same set of risks.

Three weeks after Komarov's death, Gagarin met Russayev at his family apartment, but refused to speak in any of the rooms because he was worried about bugs – listening devices buried in the walls or hidden in light fittings and telephones. The lifts and lobby areas were not safe either, so the two men trudged up and down the apartment block's echoing stairwells and along the corridors. Anything to keep moving and confuse the eavesdroppers.

The Gagarin of 1967 was very different from the optimistic and carefree young man of 1961. Komarov's death had placed an enormous burden of guilt on his shoulders. 'He told me the story about the huge research effort undertaken to try and prevent the flight,' says Russayev. 'He said the results were supposed to have been reported to the Main Man [Brezhnev]. He explained how they'd thought of me as an envoy in charge of getting the letter to the relevant offices. I told Yuri how I'd worked on it, and everything that had happened . . . He warned me, "Walls have ears." It was Yuri's idea to avoid the lifts. Somebody must have told him my apartment was bugged . . . I found out for sure when my wife woke me up at three in the morning, and we both heard a rustling behind the ventilation grille where they were installing the bug. The thought of it made me furious. How could they bug one of their own agents? I suppose that's the essence of Soviet life. There were always so many bugs around.'

At one point Gagarin said, 'I must go to see the Main Man personally. Will he see me, d'you think?'

Russayev says, 'I was amazed he could ask me this. I said, "But Yuri, you're the one who's always standing next to him on the Mausoleum. You're always chatting together, and now you're asking me if I can tell you whether or not he'll see you? I haven't even shaken the guy's hand."

' "Yes, but I never talk seriously with him. All he ever wants to do is hear dirty stories and jokes from all my foreign trips."'

Gagarin was profoundly depressed that he hadn't been able to talk properly to Brezhnev and persuade him to cancel Komarov's launch. As Russayev explains today, 'Relations between Khrushchev and Gagarin were absolutely excellent, but with Brezhnev it wasn't so good. If people don't want you, it can be hard to get through.'

Shortly before Gagarin left, the bitterness and intensity of his anger became obvious. 'I'll get through to him [Brezhnev] somehow, and if I ever find out he knew about the situation and still let everything happen, then I know exactly what I'm going to do.'

Russayev goes on, 'I don't know exactly what Yuri had in mind. Maybe a good punch in the face.'

Russayev warned Gagarin to be cautious as far as Brezhnev was concerned. 'I told him, "Talk to me first before you do anything, and I'll try to advise you. I warn you, be very careful." But I wasn't in the space department any more. I wasn't even in Moscow, so there wasn't much I could do. I don't know if Yuri ever got to see Brezhnev, and I've felt guilty ever since that I couldn't stay with Yuri to guide him.'

One story has it that Gagarin caught up with Brezhnev eventually and threw a drink in his face.

Although Gagarin grieved for Komarov, who had always been one of the ablest and most likeable cosmonauts, he remained as determined as ever to fly, and was extremely disappointed when his superiors decided to ground him from further rocket flights. Alexei Leonov explains, 'After Komarov, the State Committee decided it wasn't possible to fly Yura, because all the problems with the Soyuz had to be corrected, and it was going to take two years to redesign the vehicle.'

It was not just the slippage in the launch schedule, but renewed nervousness at the possibility of losing Gagarin to an accident, that contributed to his grounding – and there were certain military traditions to uphold. Sergei Belotserkovsky reluctantly agreed with the decision to ban the First Cosmonaut from further missions. Although he is well aware that Gagarin desperately wanted a moon flight, he says, 'The main candidate [for a possible lunar attempt] was Andrian Nikolayev. Regarding Yura, Korolev told me shortly before his death that he probably shouldn't fly any more. Yura was in a difficult situation, because he was Deputy Director of the Cosmonauts' Training Centre, and the responsibilities of that job are clearly laid out – the control and training of other cosmonauts. It's not usual for the chief of a training centre to make flights himself.'

Gagarin was very depressed by this decision, and wrote a letter to the State Committee in which he pleaded, 'I can't be prevented from flying. If I stop flying, I will have no moral rights to lead other people whose life and work are connected with flying.'

With the straight-talking wisdom of an honest working man, Gagarin's favourite hairdresser Igor Khoklov says, 'Yuri couldn't live without flying. It was his whole life. A man can't live without his trade. He can't survive.'

When the redesigned Soyuz finally flew successfully for the first time, on October 26, 1968, Gagarin's harshest critic, Georgi Beregovoi, was at the controls.

The truth behind Komarov's accident and Gagarin's grounding is only now coming to light, but most Western analysts knew by now that something was wrong with the First Cosmonaut's career. As long ago as 1982, in his ground-breaking book *Red Star in Orbit*, the American space writer James Oberg wrote:

There was Yuri, transformed before his death at thirty-four from a personable, cocky jet pilot into a demi-god to be worshipped, emulated and protected from all risk and adventure, until his own attempts to break out from the protective walls around him went just a little too far.

Gagarin diverted himself with more partying, prompting a disappointed Kamanin to note, 'Since Komarov's death, Gagarin has been dismissed from all space flights. He has undergone a new, more stormy process of personality disintegration.'

At the beginning of March 1968, the last month of Gagarin's life, a comfortable accommodation centre for cosmonauts was at last completed in Star City. Alexei Leonov remembers some hard partying, perhaps triggered by the cosmonauts' desire to block out the emotional horror of Komarov's terrible death. 'We probably met at Gagarin's apartment more often than anybody else's place. The traditions of hospitality were already established, from where we lived before, at Chkalovsky. There was this law – if you arrived late for a party you had to strip down to the waist and get into a bath of cold water and submerge your head. Even famous people had to go through this. The law was the law! Actually this was a tradition started by Yuri, and the point was that after a cold bath you were revived with a big jolt of vodka so as not to catch a cold. The trouble was, everybody started to turn up late to get their vodka.'

One distinguished guest was the architect Komarovsky, responsible for the tall tower at Moscow State University, where the first cosmonauts had been dropped down the lift-shaft back in 1960. He was welcomed with an ancient peasant gesture of hospitality, still observed by modern Russians, even those aboard the Mir space station: gifts of vital foodstuffs to protect the traveller against hunger. 'We took Komarovsky up to the top floor, where there was some bread and salt and vodka,' says Leonov. 'Then from the eleventh floor down to the tenth, where there was more bread and salt and so on, all the way down through every floor. Komarovsky, and some other famous people, they said at the end of all this, "Well, we've seen many extraordinary things in our lives, but never so much bread and salt!" Anyway, that's how we thanked the people who built our apartments.'

Certainly for Gagarin, these parties distracted him from his anxieties. Zoya recollects that when he was back home in Gzhatsk, his innermost fears occasionally surfaced:

'Yes, it's true, it was on December 5. He always came home at that time of year to see us, and to go hunting. Just as he was getting ready to leave, Mamma had some sort of anxiety, and

I remember Yura saying, "Everybody in the world asks me for something. I'm always helping complete strangers, but you never ask anything of me. You never tell me what you need." Valya and the girls [Lena and Galya] were already waiting in the car, but I had the feeling Yura didn't want to leave us. I think he was worried about something.'

12

WRECKAGE

American astronauts in the 1960s took great pride in their flying skills, and their employers at NASA gave them every opportunity to hone their skills in the air. They were assigned discretionary access to Northrop T-38 training jets, which they used as personal transports between the major NASA facilities in Texas, Florida and Alabama. These fast, lightweight planes were the space-age equivalent of company cars.

By contrast, pilots recruited into the Soviet space programme from various Air Force squadrons found to their dismay that their flying time was greatly reduced, and they were forbidden to make any solo flights, no matter how great their previous experience in the air might have been. Although the airbase at Chkalovsky near Star City provided an obvious venue for flights, very few aircraft were made available to the cosmonauts. Equipping Star City with modern jets was always a struggle, because most of the hardware had to be requisitioned from rival organizations: the Air Force in particular. Vladimir Shatalov, ex-cosmonaut and Chief of Training after Kamanin's enforced retirement in 1971, described how hard it was to obtain new jets for Star City's use:

> We have to expend a lot of nervous energy to resolve very straightforward matters. For example, we need three aircraft. It's quite obvious what they're for – but no, in order to get decisions we have to go round in circles to the Finance Ministry, the Aviation Ministry, to one appointment after another. And time goes by ... We have to become hustlers ... Is this how it should be? The

most complex space flight is simpler than all this terrestrial red tape.[1]

All aircraft pilots need to fly a minimum number of hours per year in order to maintain their qualifications. Cosmonauts at Star City who wanted to top up their conventional flying hours had to share a couple of MiG-15UTI tandem-seat trainers, which were among the most antiquated aircraft in the Soviet armoury. The first single-seat MiG fighters (with engines based on designs acquired from the Rolls-Royce company) had entered service as far back as 1947. Throughout the 1950s they were refined into one of the world's most potent combat weapons, but by the end of the next decade these old machines were no longer at their best. Communist allies abroad still purchased them in large numbers, but the domestic Air Force was switching to far more advanced fighters. Denied further flights into space after Komarov's death, Gagarin wanted to qualify in one of these newer jets, but first he had a great deal of catching up to do.

Although he was the most famous pilot in the world, he was not a particularly experienced one. Telltale clues can be discovered even to this day in the museum at Star City, where a number of Gagarin's personal effects are preserved. His pilot's log book is a much-venerated object, yet it makes disturbing reading. When he was recruited into the first cosmonaut squad at the end of 1959, his total flight time amounted to 252 hours and twenty-one minutes. Of this, only seventy-five hours had been spent as a solo MiG-15 pilot, first at Orenburg, then on station at Nikel in the Murmansk region.

For a young Air Force lieutenant starting out on his career, this was not an especially poor total, although most of the other cosmonauts in his group had logged 1,500 hours or so. If he had stayed on active duty with the Air Force, Gagarin could have built up his flying time to become a superbly skilled fighter pilot. After he was recruited for training at Star City, however, he lost this opportunity altogether. Throughout the entire period of his cosmonaut career, from 1960 to 1968, he accumulated only seventy-eight hours additional time in the air

– none of them solo. This amounted to less than ten hours per year.

On February 18, 1968, Gagarin at last received his diploma papers from the Zhukovsky Academy, greatly improving his future career prospects (on the ground at least) with a significant and hard-earned qualification. Meanwhile the position of his immediate superior at Star City, Nikolai Kamanin, was under threat because of the Soyuz accident that had killed Komarov. Although not directly responsible for the many hardware problems that contributed to the crash, Kamanin was one of the officers in authority who had sanctioned the flight in the first place, and there was a chance that his head might roll for it. There was a real possibility that Gagarin might be promoted to the rank of General, and appointed Head of Cosmonaut Training in Kamanin's place. The main worry on his mind was how to maintain the respect of the cosmonauts, a good many of whom had far more piloting experience than he did – as Beregevoi had so charmlessly pointed out.

According to an *Izvestia* journalist, Boris Konovalov:

It all worked out rather oddly. Everybody assumed that cosmonauts were pilots by profession, but they got fewer hours of flying time. When Gagarin was made Deputy Chief of Training at Star City, he made a firm stand for flying. One cosmonaut, Vladimir Shatalov, had flown every kind of jet fighter, but at Star City he was only allowed to fly a training plane with an instructor present. This was absurd.[2]

Alexei Leonov justifies Gagarin's desire to get himself – and others – back into the air. 'People were asking, "Why does he have to fly?" It was because he was Deputy Chief of Training at Star City, and in order to do that, he needed to be an impeccable pilot.' In other words, a man in the position of teaching other fliers needed to keep their respect by being a good pilot himself. Gagarin's wife Valentina hinted at the problems he faced in a 1978 interview with Yaroslav Golovanov:

He lived through some very difficult moments when the question of whether or not he was to be allowed to fly was being decided. 'And does he really need to fly at all?' someone asked. But you had to know Yura – to him, not flying would have meant not living. His passion for flying was incurable. 'Don't be upset,' I said, trying to calm him. 'How can I be in charge of training others if I don't fly myself?' he replied, much offended.[3]

By March 1968 Gagarin had not flown for five months. He turned for help to Vladimir Serugin, an experienced flier and a good teacher. As a young man, Serugin had flown 140 combat missions against the Nazis. Taking into account the late collection of confirmation signatures from his superiors, his total number of sorties probably reached 200. He shot down seventeen enemy aircraft, putting himself in the 'fighter ace' category. By the war's end he was just twenty-four years old, and a prime candidate for flying the best available new planes well into the 1960s.

In 1968 Serugin was in his late forties. Perhaps a little too old and too slow by then? It seems unlikely. As a test pilot he gained a reputation for pulling safely out of bad situations, or 'coming unscrewed' as the saying went. On March 12, 1968 he took out a newer version of the MiG, a model 21, and halted his take-off run just before he became airborne. He was convinced that something did not feel right. He taxied the plane back to the hangar and insisted that the mechanics check his engine. They found nothing wrong with it. Again Serugin took the plane to the runway, and again he turned back at the last moment. Sure enough, on closer examination the mechanics found a problem with the engine. This story suggests a flier at the peak of alertness, his instincts undimmed by early middle age.

Two weeks later, on March 27, Gagarin took off from Chkalovsky (the airbase directly alongside Star City) aboard a two-seater MiG-15UTI jet, with Serugin in the back seat acting as his instructor. The purpose of the flight was to prepare Gagarin for qualifying in a more modern MiG-17, so that he could leave the older plane behind once and for all.

Valentina was in hospital undergoing an appendectomy, and Gagarin planned to visit her later on, at the end of his day's work.

At 7.30 in the evening, Taissia Serugina started to worry because her husband Vladimir was not yet home. As she remembers, 'I was waiting for the whole night. I called his air regiment, and every time they said, "He's not available, but everything's in order. He's busy with his work." No one told me anything. I didn't sleep, and I left the house next morning for work. Then they notified me that there had been a problem at the airfield, but I didn't quite believe it. I thought if anything serious had happened to my husband, they would have told me yesterday . . . Suddenly my daughter ran up to me. "Mother!" she shouted, and there were tears in her eyes. "Father's dead!" I don't remember much after that.'

Alexei Leonov was one of only a few cosmonauts to have embraced helicopter flying as a worthwhile discipline. He was involved in the testing of possible lunar landing manoeuvres using adapted helicopters as crude vertical-descent simulators. On the morning of March 27, he was leading a group of cosmonauts through a parachute training run from the Kerzatch airfield, thirteen kilometres from Serugin's and Gagarin's base at Chkalovsky. He piloted a large helicopter through the deteriorating weather, trying to find a break in the clouds so that he could release his jumpers.

The cloud base was down to 450 metres and visibility was appalling. Rain and wet snowflakes thudded against the cockpit canopy. Leonov managed to release his first parachute team into the air, but the visibility was closing in fast. The local air-traffic controllers told him that the weather was not going to improve, so he took the helicopter back to Kerzatch with half his parachute team still aboard. 'Moments after we had landed, we heard two explosions – an explosion and a bang that accompanies a supersonic shockwave. We wondered: what was it? An explosion or a bang? I said it was probably both – that the events were somehow linked. And these two sounds were just over one second apart.'

Chkalovsky was thirteen kilometres away and the sounds were muffled by the damp weather, but even at that distance they were distinguishable. Leonov became increasingly concerned. He knew perfectly well that Gagarin was flying today. On his own authority, he flew the helicopter to Chkalovsky, despite the poor weather. All the way there he monitored the controllers calling Gagarin's code number, 625, on the radio link. As soon as Leonov touched down at Chkalovsky, a regimental commander came up to him and said, 'The fuel in Yuri's plane should have run out forty-five minutes ago, but he's not returned to the airfield.'

Leonov decided he had better report his unpleasant theory. 'I went to the Flight Control Office and Nikolai Kamanin was there. I told him, "You might think it's strange to say this, but I heard an explosion and a supersonic bang." I gave an estimation of the [compass bearing] I thought the sounds had come from.'

A search helicopter was despatched to overfly the area where Gagarin's plane had last been spotted on radar, ninety-six kilometres north-east of Moscow. The pilot flew low over the ground and discovered an area of woodland with a bare black patch of scattered earth venting some steam, but visibility was still poor and he could not be sure that this was actually a wreck site. According to Leonov, 'The search pilot thought the steam might be a natural phenomenon of some kind. He was ordered to land his helicopter and inspect the site on foot. Because of all the trees there was no obvious opportunity to put the helicopter down, so the pilot flew to the nearest open land, near a church, and settled there.' Apparently he waded for an hour through thick snow, a metre deep in places, to get into the woodlands where he had seen the smoke. When he had found what there was to find, he struggled back to the helicopter and made his report by radio. There was a large crater, he said, and the earth from within it had been thrown outwards across a wide area. Some of the trees at the perimeter were broken, and many small pieces of twisted metal lay all over the site. Clearly this was an aircraft accident, but there was no obvious sign of a central piece of wreckage in the crater: a fuselage, for instance, or a main engine section.

Gagarin and Serugin had lost contact with Chkalovsky traffic control at 10.31 in the morning. By the time the helicopter pilot had waded in and out of the wreck sight, made his report and called for a properly equipped rescue team, it was about 4.30 in the afternoon. The grey winter light, already poor, was fading fast. The search team arrived with powerful torches, but they were of little use in the winter darkness. By evenfall the searchers had identified what appeared to be tatters of Vladimir Serugin's clothing, and Gagarin's map case, but they had found no obvious trace of either man's body, nor of the main sections of the aircraft. 'Throughout the night two battalions of soldiers searched the forest, but they didn't find anything,' Leonov explains. 'And on the next day, while we were digging deeper into the crater, we found pieces of Gagarin's flying jacket. It became clear that both of them were still in here somewhere. They didn't eject.'

The front end of the plane had been rammed with great force several metres into the hard ground, by the sheer momentum of the heavy engine block. The recovery team had to try and dig the cockpit out of the hard-frozen earth. They found that it was utterly smashed, and the two men's bodies inside were severely mangled. To their great distress, the rescuers spent many hours retrieving fingers, toes, pieces of ribcage and skull from the crater, the surrounding woodlands and even the trees – some of these had to be cut down once they knew what to look for. It became clear that the plane's impact with the trees had caused terrible damage to the cockpit, even before the final impact on the ground had crushed it once and for all.

Meanwhile Gagarin's personal driver Fyodor Dyemchuk, who had driven him to Chkalovsky that morning, was quietly waiting for the MiG to return so that he could get his passenger back to central Moscow, to see Valya in the Kuntsevo hospital in the evening. 'At approximately eleven o'clock [that morning] all of us learned that his radio link was lost. Everyone assumed his transmitter was out of order or something like that.' But the mood darkened once the search party was ordered later that day. 'We were told that a crash site had been found and we were under orders to be ready at eight o'clock in the evening. We formed a team, picked up some equipment and went to that place. There was a lot of snow, and the ground was difficult, so it took us most of the night to drive through and reach the crash site. Of course everyone was upset. Everyone felt it. The most horrible thing was the uncertainty.'

At first light next morning, the extent of the crash became clear. Dyemchuk was closely involved in the search to recover every scrap of wreckage, no matter how small or seemingly insignificant. 'The only large pieces left were the engine, some landing gear, and one wing. The rest was scattered over the entire forest by the force of impact and the explosion. We were walking through the snow. You walk and see a hole in the snow, and you dip your hand in and pull out a piece of flesh or a piece of bone. Sometimes a finger. Those were very dark days.'

Dyemchuk's worst moment actually came two days after the crash, when he was driving a distraught Valentina Gagarina away from the hospital after her operation. Thoughtlessly he let slip some comment or other about recovering Gagarin's body. 'She was hysterical. She didn't know. She thought he was found intact, or at least that the majority of body parts were found. Of course men understand very well what happens in an explosion, but how could women know about these things? She didn't realize they were blown to pieces. Because of my naïvety, I told her. Perhaps a bitter truth is better than a sweet lie.'

Under conditions of the greatest security, Leonov, Kamanin and other colleagues were asked to attempt an identification of the two dead pilots' body fragments. Leonov says, 'When they showed me part of a neck, I said, "That is Gagarin." Why? Because of a birthmark. On Saturday we were at the barber's shop at the Yusnost Hotel. There was a barber, Igor Khoklov, who liked Yura very much, and he always cut his hair. I saw the birthmark, about three millimetres across, and I said, "Igor, be careful. Don't cut it off." So I knew when I saw it that we could stop searching. We wouldn't find Gagarin out there somewhere. He was here.'

Meanwhile, one of the most intensive air-accident investigations in Soviet history was initiated. Despite the very wide scattering of wreckage, 95 per cent of the MiG-15 was recovered for analysis over the next fortnight. Even while this painstaking recovery was being carried out, fragments of heart and muscle tissue from the pilots' shattered bodies were sent off for chemical analysis.

A standard sequence of biochemical tests was performed on the remains of all Soviet military pilots killed in accidents. Lactic-acid levels in muscle tissues yielded clues to a pilot's physical condition at the time of a crash. High levels of acid suggested tightly flexed muscles and a thoroughly alert pilot. Low levels indicated a relaxed state, perhaps a result of unconsciousness brought on by extreme g-forces. In such a case the accident investigation was fairly straightforward. The pilot could be deemed responsible for the crash, but his honour was protected. Another possibility, suggested by intermediate lactic-acid levels, was that tiredness might have caused the pilot's attention to wander; in which case the investigation was widened to include his overall workload and career. The worst possibility was alcohol. If a pilot was found to have been drunk in charge of his aircraft, his reputation could not possibly be redeemed. The chemical tests searched for traces of alcohol as well as lactic acid.

Immediately after Gagarin's and Serugin's crash a rumour spread that they had indeed been drunk. This story is still put about today.[4] They went to a fiftieth birthday party for a colleague the night before their flight, and partied hard and long. Taissia Serugina utterly rejects this idea. 'The night before the flight my husband went to bed at ten. I asked him, "Why are you going to bed so early?" He said, "Tomorrow I have to test Yura, so I want to be in good shape." In the morning he left for work in a good mood. He said, "It'll be a good day today." But a tragedy occurred.'

Taissia admits that a party did take place prior to the crash, but two nights before. 'On Monday there was a celebration in Star City for a colleague's fiftieth birthday. On Tuesday my husband was working as normal. On Wednesday Yura was due to fly. That's why, on Tuesday evening, my husband told me he'd go to bed earlier.' Taissia blames the rumours of drunkenness on Serugin's immediate superior at the Chkalovsky airbase, General

Kuznetsov, who treated Serugin with considerable discourtesy throughout their working relationship. 'He would summon my husband to his office, then keep him waiting outside. Finally my husband would become exasperated. He would arrive, only to find that the man wouldn't see him, so he'd turn around and drive back to the airfield.'

The problem between the two men appears to have been rivalry for rank within the Chkalovsky airbase. Gagarin and Serugin were very good friends, and Taissia Serugina is convinced that General Kuznetsov resented her husband's closeness with the First Cosmonaut. 'Yura said to my husband, "Don't pay any attention to Kuznetsov, because very soon I'll be Chief of Training and everything will work out." Afterwards, Kuznetsov said that my husband was ill during that last flight, and he had a sick stomach or an ulcer. Never in his life did he complain about any illness. To say such foul things is absolute dishonesty.'

If Kuznetsov was using the phrases 'ulcer' or 'sick stomach' to suggest a hangover, then the hard evidence supports Taissia Serugina's side of the argument. Samples from Gagarin's and Serugin's remains were sent to several institutes, and all of them reported similar results. Lactic-acid levels in the muscle tissues of both men were high, indicating that they were fully conscious and alert at the time of the crash. In fact, the levels suggested an intense physical battle with the MiG's control yokes (what we would call 'joysticks'). Meanwhile the alcohol levels were not found to be significant.

The aircraft wreckage revealed other clues. The yokes in the front and rear cockpit compartments were positioned as they should have been by pilots attempting to control a wayward aircraft. In theory, the crash could have dislodged the yokes entirely by chance, but the foot pedals also appeared to be in the right positions. Likewise for the throttle levers and flap controls. Despite the extreme damage to most of the cockpit's mechanical components, there was strong evidence to suggest that both pilots fought hard to save the plane from a catastrophic spin. What's more, they seemed to have been trying all the right manoeuvres, aiming for a subtle 20-degree tilt of the aircraft, rather than simply hauling at the yokes in thoughtless panic.[5]

Alexei Leonov was a member of the crash investigation team. As he points out, 'The plane was not in a dive when it hit the ground, but coming out of it. The plane didn't crash nose-first, but almost flat on its belly.' Its downward momentum was still sufficient to drive the engine block several metres into the ice-hardened earth; but the 'pancake' impact suggested that the two pilots may have been heart-breakingly close to levelling out when they ran out of sky.

The central question was: why had the plane lost control in the first place? Obviously it had not collided with another plane, otherwise it would have disintegrated in mid-air, scattering wreckage across a much larger area of ground. And there would have been signs of wreckage of the plane which it had collided with.

The crash seemed a puzzle. The investigators turned for answers to the service record of Gagarin's and Serugin's MiG. Perhaps the ageing jet had somehow failed, or lost power? The commission noted several concerns:

> Shortcomings of the equipment and procedures used in the flight:
> 1 The MiG-15UTI aircraft was old, produced in 1956 and subject to two major overhauls. The residual service life of the structure was down to 30 per cent.
> 2 The engine, DA-450, was also produced in 1956 and subject to four overhauls. Residual service life was 30 per cent.
> 3 Installed on the aircraft were two 26-litre external tanks which were aerodynamically poor, reducing allowable g-loads by a factor of three.
> 4 The crew ejection system required the instructor to be the first to eject.
> 5 The height-to-ground altimeter was faulty.[6]

Gagarin and Serugin flew with expendable drop-away fuel pods under the wings. Leonov says, 'There was always a drawback with this configuration. The dynamic design of the fuel tanks reduced the safety parameters of a flight, such as the angle of attack, the angles of sliding, and g-forces.' Drop-tanks were not required to sustain a brief training sortie directly over home base. The tanks' usual purpose was to supply a combat MiG with enough fuel to reach foreign enemy territory. When it reached its intended war zone and began to fight, the MiG was supposed to drop the depleted tanks in order to regain maximum nimbleness and speed. On March 27 a pair of tanks was installed on Gagarin's training plane so as to familiarize him with the extra care he would have to take while they were in place. They should not have presented any special problems. All MiG pilots understood the strict regulations forbidding them to attempt simulated combat manoeuvres with the tanks attached.

In general, the MiG-15UTI was a rugged machine that allowed its cadet students plenty of margin for error – tanks or no tanks. Everyone called the UTI configuration 'mother', because so many thousands of pilots had learned to fly aboard these two-seater machines. Despite doubts about the age of Gagarin's jet, and its damning record of overhauls, the wreckage did not suggest that any structural failures had occurred prior to the crash itself. So why did an intact and fully functioning MiG suddenly fall to earth? Seeking an explanation, the commission looked into the weather reports for the day of the crash:

> Difficult weather conditions became gradually worse, as evidenced by ring-shaped pressure contours on the meteorological charts, and eye-witness accounts. Inaccurate weather information was given [to the pilots] during pre-flight preparation because the sortie of the weather reconnaissance planes was delayed.[7]

Apparently Serugin was misinformed that the cloud base was at 1,000 metres, when in fact it was down to 450 metres. A slight fault in the MiG's instrumentation prevented its altimeter from responding accurately if the plane was in a dive. Serugin may have descended through the cloud layer thinking that he had twice the height margin over the ground than was actually the case. The difference in flight time would not have been more than a few seconds at most, but it could have been crucial.

Were those few seconds of visibility beneath the cloud layer sufficient for Serugin to see that he was running out of altitude? If so, then why didn't he order an emergency ejection before the plane actually hit the ground? According to Leonov, the minimum safe height for ejection from a MiG was as low as 200 metres, but the belly-first pattern of the crash suggested that Serugin may have thought he was on the verge of pulling up safely, which could be why he did not order an ejection; and, as everyone knew full well, if he had considered bailing out, the escape procedure itself might have caused some difficulty. Two decades after the crash investigation, Igor Kacharovsky, an expert aircraft engineer, wrote to Sergei Belotserkovsky with his observations:

> As a rule the MiG-15UTI is flown by a cadet and a flying instructor. The instructor sits in the rear seat. The front seat is in the same position as for a single-seat combat MiG-15. The order of ejection is as follows: the first to eject is the instructor from the rear seat. The second is the pilot from the forward seat.
>
> If the forward pilot ejects first, the gas jets from his ejection mechanism will interfere with the rear compartment, thus making ejection from it impossible. Instead of finding a better technical solution, the designers made a 'methodological' decision without thinking about the consequences. The instructor had to be first to eject, which is contrary to common ethical standards.[8]

Kacharovsky's point was that no instructor worth his salt wanted to leave a less experienced pilot to fend for himself in a stricken plane. The instructor would be honour-bound to get his pupil out of the plane first, before tending to his own survival. The arrangement aboard the MiG-15UTI flouted this honourable tradition. Though he can offer no proof that an ejection was ever attempted, or even considered, Kacharovsky suggested the following terrible scenario:

> It is easy to imagine the situation. Serugin, as crew leader, ordered Gagarin to eject, but Gagarin understands that saving his own life exposes the life of his teacher and friend to danger. Each man thinks of the other.

Kacharovsky imagined the two men arguing the position, wasting valuable seconds until it was too late and they hit the ground. In fact, this scenario is more emotional than strictly logical. The MiG-15UTI designers had little choice in the ejection sequence. All two-seater jets around the world employ the same sequence, for a very simple reason. If the pilot in front ejects first, then the plane moves fractionally forward beneath him as he flies upwards. In the brief fractions of a second while he is still close to the top of the plane, the rear cockpit position passes directly under his ejection seat, so the rear pilot's escape route is blocked for valuable fractions of a second. What's more, the rocket blast from the first seat burns through the rear canopy, putting the other pilot's life in grave danger. However, if the rear pilot ejects first, the man in front can then fire his own seat safely, because his rocket exhausts will trail over an empty rear cockpit position.[9]

There is no moral shame attached to this. The safety margin between the separate ejections is less than half a second. The supervising officer in the back seat issues the order to eject and immediately exits the aircraft. The cadet in the front seat responds so quickly afterwards that the difference in timing is barely significant.

Much more significant was the fact that the cockpit canopy frame was found among the wreckage. In a modern jet fighter a pilot in danger pulls a simple lever on his seat, and the elaborate ejection mechanisms take care of everything else, including the removal of the canopy. If worst comes to worst and the canopy does not come away properly, then a web of explosive wires built into the plexiglass shatters it, so that the seat can simply punch its way out. On the old MiG, however, a separate mechanical lever on the pilot's left side had to be pulled first in order to dispose of the canopy. Only then could he eject. Obviously neither pilot had pulled the canopy jettison lever.

But the canopy frame in the wreckage did not have much plexiglass left within it. Most of the transparent material was shattered, and only a very small proportion was recovered at the wreck site. This was the only physical evidence in the entire investigation that directly suggested a mid-air collision of some kind. If the MiG had smashed into another plane, there would have been much more in-flight damage than merely the shattering of the canopy. The missing glass was suggestive of a grazing impact with a bird, or with the suspended instrument package of a stray weather balloon – and this is the principal explanation for the crash that the commission eventually settled on, based on the one solid piece of evidence: that the plexiglass was missing. Serugin and Gagarin lost control of their plane when the canopy shattered, and did not quite manage to recover.

The KGB conducted a parallel investigation, not just alongside the Air Force and the official commission members but against them. Their report also focused on the simplest possible explanation, as adopted by the commission, based on the shattered cockpit canopy. One of the KGB investigators, Nikolai Rubkin, today a 'State security expert', knows all aspects of the security service's relationship to the early space effort. He is one of the few people who can gain access to the voluminous original report, stashed way even now in the bowels of the Lubyanka. He says, 'The missing plexiglass in the canopy meant that something must have hit the cockpit before the crash. A bird strike would tend to hit the front of the canopy, not the top. An impact with an aircraft would have created much more damage. The missing glass is more consistent with an impact against the suspended instrument package of a weather balloon.' So could the commission's findings actually be correct? 'The only indisputable fact is that the cockpit canopy glass was broken before the plane hit the ground,' Rubkin says, carefully. 'Everything else is guesswork. Only Gagarin and Serugin could tell us the truth about what really happened that day.'

221

Rubkin puts the investigation's politics into broad perspective: 'There were several sub-commissions investigating different areas. One of them dealt with the aircraft's maintenance, another with pilot preparation, a third with the fuelling and tank installation, and a fourth examined all the medical matters. Finally there was another looking into any possibility of sabotage, or a revenge plot. That last was very much the KGB's responsibility at the time.' The problem – as so often with a high-profile and politically sensitive investigation – was that the five sub-commission teams did not communicate with each other. 'Since there were several major institutions responsible for all these various areas, and the KGB had its own departments, the sub-commissions' documentation was never assembled as one coherent package for the main commission. The reason was that too many interested parties worked for institutions that might have been found responsible for the crash. Certain people, whether we like it or not, adjusted the facts to save their honour. I found a report from General Mikoyan, the famous man who designed the MiG in the first place, saying that he was completely dissatisfied with the way the investigation was carried out.'

Alexei Leonov and Sergei Belotserkovsky also remained thoroughly dissatisfied with the commission's work. They thought the weather-balloon theory was completely wrong. Leonov thinks he knows exactly what happened that day. 'Another plane passed very close to Gagarin's and Serugin's MiG in the clouds, coming within ten, fifteen, twenty metres. The vortex [backwash] from the other plane turned the MiG upside-down and caused the loss of control and the crash.'

Leonov's theory about aerodynamic interference from another plane provides a credible explanation for the disaster, except that such an ordinary problem should have been survivable. If backwash was a factor on March 27, Serugin should have been able to stabilize the MiG without too much difficulty. Major-General Yuri Khulikov, a former Air Force Chief of Flight Security Services, points out that the MiG-15 had been intensively flight-tested under simulated backwash conditions. Given a reasonable altitude for safety, any averagely experienced pilot should have been able to regain control. In January 1996 Khulikov gave an interview to *Moscow News*, in which he focused rather harshly on 'pilot error' as an explanation for the crash. 'Even if Gagarin and Serugin got into a vortex stream, the MiG should have been recoverable. Such a vortex doesn't much affect the engine. I'd like to point out, this conclusion was reached after a very stringent series of tests . . . Gagarin wasn't at all prepared for such conditions . . . You must understand what the name "Gagarin" meant in our country at that time. It was a symbol of the victory of socialism in space. It seems that the First Cosmonaut couldn't be capable of making mistakes.'[10]

But Khulikov has an axe to grind, since his loyalties lie with the original members of the 1968 investigating commission and those senior officers responsible for general air-traffic control at that time. Notably he forgot to mention that the MiG-15 and many other planes used in vortex recovery tests had never been fitted with drop-tanks because – and this is where the military logic goes round in circles – it was forbidden to fly drop-tanks in such extreme manoeuvres. Quite simply it never occurred to anyone to test MiG-15 training craft under the most severe flying conditions with the tanks attached, because it would have been much too dangerous, even for the most experienced test pilot.

Clearly a backwash hitting a MiG with drop-tanks (as flown by Gagarin and Serugin) would have been more of a hazard than Major-General Khulikov likes to admit.

Alexei Leonov goes much further, insisting that it was not just ordinary backwash from another MiG, but a powerful supersonic shockwave from a brand-new, high-performance fighter that slammed into Gagarin's and Serugin's plane like a solid brick wall.

Leonov was always firmly convinced that the two bangs he had heard after he landed his helicopter at Kerzatch were made by two entirely different phenomena. The MiG-15UTI was fast, but far from supersonic. The distant bangs may have sounded faint from where he was standing at the time, but he was sure they were caused by an explosion and an additional supersonic boom.

Therefore another, and much faster, aircraft must have entered the same airspace at the wrong moment. But when Leonov tried to persuade his fellow investigators to explore this theory, 'all my attempts were stopped by some invisible wall. I understand that a Deputy Chief Commander was appointed to the accident commission. He was also in charge of the traffic control for that region, and he could have been responsible [for events on March 27], but he didn't pay attention to them in his report. It could have been problematic.'

Leonov was unhappy about such obstruction. He was sure that the supersonic boom had not been a figment of his imagination. Eyewitnesses on the ground, near the crash zone, contributed some powerful supporting evidence, which again was not included in the final report. 'Apart from the fact that I heard the sounds myself, three local dwellers were questioned separately. All of them said they'd seen smoke and fire coming out of a plane's tail. Then it went up into the clouds. So it was a reversed process. Gagarin fell down to earth, but this other plane went upwards at great speed.' The witnesses were shown aircraft identification charts, and all of them immediately picked out the distinctive outline of a new Sukhoi SU-11 supersonic jet, which looked nothing like an old MiG-15. 'We knew that SU-11s could be in that area, but they were supposed to fly above 10,000 metres,' says Leonov.

The 'smoke and fire' coming from the mysterious plane's rear end were sharply suggestive of an afterburner at full thrust. The SU-11 included an afterburner, which was a relatively new piece of technology: a supercharger wherein the jet exhaust was re-ignited for extra thrust, particularly when the plane was pushing towards supersonic speed and beyond. At full thrust the SU-11 could achieve nearly twice the speed of sound. The antiquated subsonic MiG-15 did not have an afterburner and its exhaust stream was not noticeably fiery.

The evidence for this mysterious second aircraft in the area was supported by one of the air-traffic controllers on duty that day, Vyacheslav Bykovsky, who told the commission that he had seen two other target blips on his radar screen, one of which was approaching from the east. Apparently that signal continued to register on his screen for at least two minutes after Gagarin had crashed. In fact, the timing of the crash was hard to define. Seismometers in Moscow registered a signal at 10.31 in the morning, consistent with an aircraft impact, but Bykovsky says, 'To this day I don't believe Gagarin fell at that time, because we lost contact with him on the radar at forty-one minutes past, not thirty-one.' Then he contradicts himself, saying that the MiG's chronometer was found among the wreckage, jammed at 10.31. 'Who knows what all this means? There are so many possibilities. Maybe the people in Moscow recorded some other shock before the crash. I don't know. I went to Star City a year after Gagarin's death, and the tour guide said he died at 10.41. A year afterwards they said he died at 10.31. There's a big difference.'

Immediately after the crash, Bykovsky and the other controllers in his station were placed under security, and their evidence at the time was carefully filtered. Today he says, 'There were two other planes in the area. We knew about them. The generals on the commission gathered us all together and we explained to them what we'd seen, but we were segregated and we didn't work again for more than a week. People were questioned about the other plane [the plane which may have interfered with Gagarin's and Serugin's flight] and many said they'd seen it.'

As Bykovsky demonstrates, the testimony about radar signals is complex and ambiguous. He readily admits that the tracking equipment was unable to keep simultaneous tabs on both the positions and altitudes of nearby aircraft. 'Either the blips on the screen appear or they disappear. If a plane changes altitude, it disappears for ten seconds, so the signal on a radar screen isn't always constant. At forty kilometres' distance from the airbase the signals disappear altogether.'

Leonov says that Bykovsky's report of at least one – and possibly two – additional target on his radar screen was discounted by the commission. 'It was attributed to his lack of experience. They took him away somewhere, and I don't know exactly what happened to him. In any case, none of this appeared in the subsequent documentation. The fact that I provided this information [about the two bangs] and spoke to people who saw the other plane – this wasn't enough for the commission. That's why no one knows about the other plane, except its pilot and his conscience.'

In fact, a second MiG pilot, Andrei Koloshov, emerged from obscurity in April 1995 to admit that he was indeed flying in the area at the time. In the journal *Argumenti i Fakti* (Arguments and Facts) he said, 'The cause of Gagarin's death was that he was reckless in taking an unjustifiable risk. He and Serugin deviated from their proper flight pattern.'[11] Koloshov suggested that the two men agreed to fly away from their designated zone in search of clearer weather, so that they could at least try out some basic manoeuvres. He presented absolutely no evidence to support his theory. Perhaps he had a guilty conscience. The original traffic-control voice tapes (finally unearthed by Leonov and Belotserkovsky in 1986 after a long battle with the authorities) show that, far from flying recklessly, Serugin had cut an intended 20-minute training session down to five minutes because of the poor weather. Bykovsky remembers without prompting that the last voice communication from Serugin was to say that 'their job was done. He told us everything he was doing. He had completed the training task and he asked for permission to come out of [the current flight zone]. Then the radio link was lost.'

Koloshov's charge of recklessness on Serugin's part seems unjust, but today Leonov is not concerned with the MiG pilot's ungenerous testimony, because he is still convinced that this pilot and his subsonic MiG-15 were utterly irrelevant and had nothing to do with Gagarin's death. 'No MiG-15 could have made the supersonic bang that I heard that morning.' Leonov and Belotserkovsky still assert that a supersonic Sukhoi SU-11, never firmly identified in the confused radar data, was the true culprit – in the air, at least.

Leonov is generous towards the SU-11's pilot, whoever he may have been. 'If he'd been identified at the time, he'd have been torn to pieces by an angry crowd. On the one hand, they should have released this information; but on the other, if we think about it wisely, perhaps not. It wouldn't remedy anything.' It was not so much a single pilot but 'the whole system that allowed for Gagarin's death. The entire system. You can't take the entire system to court. You can judge it morally, but you can't punish it.'

The 'system' did not want to be judged or punished. In all, the commission's report accumulated twenty-nine thick volumes of technical data, but the mere appearance of fact-gathering was not the same as making a fair analysis of the causes. The 1968 commission report's central judgement was deliberately vague and simplistic: 'an aggregate of causes'. The main thesis – the grazing impact with a weather balloon – suited everyone because it was the most innocent. No one was to blame – at least, no one on the ground.

One of the commission's hardest-working investigators, Igor Rubstov, supported Leonov's and Belotserkovsky's theory that a supersonic aircraft had come close to colliding with Gagarin's and Serugin's MiG. As the commission moved ever further away from this difficult territory, Rubstov gathered all his courage and went to the KGB headquarters in the Lubyanka to argue his case. 'I can't say I felt too confident in that particular building.' He met Colonel Dugin of the KGB, who demanded to know why he was insisting on the collision theory. Rubstov bluffed in classic Russian style. 'I said that if the commission failed to investigate [the near-collision scenario], people might think there was something to hide. It was better to investigate it properly to demonstrate that it was not the true version of events.' Colonel Drugin was unimpressed. On his desk there was a slim folder, which he now opened. It turned out to be Rubstov's personal file. 'You don't honour discipline very much, do you?' the Colonel said.

Rubstov knew that he was referring to an incident during the war, when an aviation unit at Stalingrad had retreated, quite justifiably, to a safer position under intense German attack. For twenty years and more, no particular inference had been drawn from this event. Drugin now implied that he could use this ancient story as proof of Rubstov's cowardice, simply because he had been a member of the retreating unit. Drugin did not have to make this threat in so many words. He merely opened the personal file at the relevant page so that Rubstov could catch a glimpse of its contents, then asked him to reconsider his ideas about the near-collision scenario. 'Later on, this version was not confirmed,' Rubstov forlornly admits.

Leonov and his closest colleagues wanted to learn the truth about the crash. It took them the best part of two decades, but in 1986 Belotserkovsky lobbied successfully for a new commission of inquiry. He gained access to the secret investigation documents and original supporting materials, including unedited voice tapes of the air-to-ground dialogue. Meanwhile Leonov was amazed to find that documents supposedly written by him in 1968 as part of the original commission of inquiry were in someone else's handwriting. 'They were rewritten, and certain effects were falsified.' Leonov's accusation comes as no great surprise to security expert Nikolai Rubkin. 'I can't exclude that possibility. We've never had any problem in our country finding people to forge signatures. There have always been plenty of ranch-hands skilled in this kind of art.'

Belotserkovsky discovered that all the radar operators were hopelessly confused at the time of the crash. 'First of all, I noticed that the tapes of the conversation between the flight controller and Gagarin's plane contain a curious moment. The thing is, the controller was still calling out Gagarin's call sign, six-two-five, after his plane had already crashed. The controller's voice was perfectly calm. He wasn't nervous. But from minute forty-two of the tape, he did show some nervousness. That was about twelve minutes after the crash.' Belotserkovsky was suspicious about the long delay in the controller's reactions. Even allowing for the sluggish response of the radar equipment, the MiG-15's blip must have dropped off the screens eventually, as the actual aircraft plunged towards the ground, but it took a full twelve minutes for the controllers to realize that anything was wrong. This may explain the ten-minute discrepancy in the timing of the crash that Bykovsky describes, with some discomfort, in his interview.

Belotserkovsky found many other flaws in the ground-control procedures, and in the original commission's fudged report. The standard practice back in 1968 was to make photographic records of traffic-control radar screens at set intervals. An automatic system of cameras was built into the consoles to do this, but on March 27 the cameras at Chkalovsky were not working, so the controllers resorted to a crude back-up recording system. They placed pre-cut sheets of tracing paper across their radar screens and marked the positions of the various targets at intervals. Belotserkovsky found the old and faded sheets tucked discreetly into a folder marked 'secondary material', as if to disguise their importance. 'There was a whole spectrum of conditions that we did not manage to cover while working on the original commission. We agreed that two particular lines of evidence, the voice tapes and the tracing-paper sheets, showed that the traffic controller was talking to a different plane, which he mistook for Gagarin's. Most likely, Gagarin's plane got so close to the other plane that they appeared for a moment on the radar screen as a single target. When Gagarin's plane went into a spin, the other one was still on the screen.'

In the poor weather, and with no warnings from ground controllers, the crew of the other jet may not even have been aware of the near-miss. But some retired Sukhoi SU-11 veteran out there may well be keeping his head down today.

Gagarin's death was shameful not just because of the loss of a national hero in muddled circumstances, but because of the dangerous flaws revealed in the Soviet military technology of his time. Obviously their radar systems were not capable of simultaneous mapping of aircraft heights and positions, nor of positively identifying one target from another. The implications of this were highly alarming. In theory, a foreign jet simulating approximately the usual routines and flight patterns of Soviet aircraft could have flown close to an airbase, or some other military target, without clearly being identified as a potential enemy. In all likelihood Gary Power's U-2 spy craft, shot down in May 1960, was identified as hostile only because its flight path was noticeably different from the expected routes of other Soviet aircraft that day.

EPILOGUE

There have always been rumours that Yuri Gagarin was murdered by Leonid Brezhnev's administration. Journalists, friends and relatives still talk of dark plots, although their anger is more metaphorical than literal. There is no real evidence to suggest that Gagarin's crash was anything other than an accident. Incompetence and poor administration at many levels certainly contributed to his death, but deliberate malice seems unlikely. The real crime, at least as far as Gagarin's family was concerned, was that the authorities told them so little of the truth. 'My parents weren't sure what to believe,' says Valentin. 'We thought Yura's death was ordered by Brezhnev. When Yura was at official visits with him, nobody paid any attention to Brezhnev, and he hated it when people didn't listen to him. Brezhnev wanted people to pay attention to him, and nobody else . . . There are no accidents in life, only causes that lead to accidents. I don't believe in coincidences, either. It was a set-up, right to the last minute.'

The last time Valentin saw his brother was on February 25, 1968, a few days after Yuri had received his diploma. Some journalists spoiled the mood that evening when they arrived, uninvited, at Gagarin's Moscow apartment. 'They rang the bell, I opened the door a little bit, and they pushed their way in,' says Valentin. 'What the hell could I do? Yura said they were parasites, and he couldn't even relax at home. They started to take pictures, and one correspondent noticed Yura's new Japanese camera, and said, "I will give you my camera, you give me yours, and I'll pay the difference." Yura turned to Valya and said, 'Let's give him the money instead, so he won't ask that question again." The journalist was very ashamed after that.'

Gagarin's sister Zoya also tells a bitter story. 'The last time we saw Yura was at his graduation on February 18, where he received his diploma papers from the Zhukovsky Academy, along with Gherman Stepanovich Titov. Yura was very happy to receive the diploma after so much hard work. After that, we only heard about his death five weeks later on the radio. We weren't given any advice, we weren't warned in advance. We weren't told anything at all. I felt very sick, and so did Mamma. The doctors had to give us endless injections to calm us down . . . There was never any precise official information about the cause of Yura's death, just guesses and rumours, all the worst things you could posibly think of. Someone helped him to die, that's my feeling.'

Zoya recalls the funeral arrangements with a grimace of discomfort. 'We sat for two days in the House of the Soviet Army, and the endless funeral music banged away in our heads. We thought we'd go insane. People were walking, walking, walking through to say goodbye. They came from everywhere, an endless procession of them. There were such long queues, the guards had to block the entrance for a while. It was terrible.'

As was the custom, Gagarin's mother wanted to see her son for one last time before committing his body to the crematorium's flames. Valentin describes the worst moment. 'We wanted to open the coffin, but the head of the funeral team wouldn't allow it. Mamma and Zoya started to argue with him, and everybody was shouting. Finally he let them do whatever they wanted. They pulled off the red velvet drape and opened the coffin, and inside there were human remains in a plastic bag. It was just about possible to recognize some of them. Yura's nose was in place, but his cheek was torn off. Somebody told me later that Serugin in his coffin looked just as bad. Well, we looked, and then we closed the coffin. The music started to play, and the coffin moved slowly into the furnace. The next day, at the official funeral, Yura's ashes were put into the Kremlin wall. And that was it.'

Zoya says that her mother Anna took her son's death very hard, and a peculiar cruelty of history prevented her from finding any peace. 'Usually people have a chance to bury their loved ones, and then they calm down as time passes, but every day Mamma was reminded of it, because Yura was so famous. People were always turning up from all over the Soviet Union to pay their respects and see our family home. Mamma lived until she was eighty, and I often wonder how she got through it. She suffered more than the rest of us, I'm sure.'

Many of Gagarin's friends and colleagues visited his parents in Gzhatsk to express their sympathy. Sergei Belotserkovsky recalls, 'During my last meeting with Gagarin's mother, when we were alone, she asked me all of a sudden, "Was Yura killed?" I was stunned. "What makes you think so?" I asked, and she said that Yuri had told her once, "Mother, I'm very afraid." She said she hadn't understood what he'd meant by that, but it troubled her.'

Belotserkovsky offers his own interpretation. 'I don't think Gagarin feared for his life. It was a different kind of fear – the fear we all shared in those days, our fear of society, and of the world in which we lived. Letters – dreadful letters – were pouring into Gagarin's office. All the distresses and problems in society impinged on him. He carried an immense load on his shoulders . . . One could sense his anxiety and tension. He was an emotional person, and he felt upset when he couldn't help . . . He didn't fit into the lifestyle of the Party élite and the higher levels in the Brezhnev system. He was alien to them, so they rejected him. There were attempts to tame him, to buy him off, but he wouldn't succumb. He was too honest, too self-willed and independent.'

Belotserkovsky, Leonov, Titov and others were welcome at the family house, but other less familiar visitors inadvertently added to the family's emotional turmoil. From the moment of Gagarin's space flight in 1961 to his death in 1968, his father Alexei and younger brother Boris were approached almost every day by people wanting them to pass on their requests to the First Cosmonaut; or by people simply seeking the thrill of meeting any available members of this famous family. After Gagarin's death, the strangers still came; and over time both Alexei and Boris became inadvertent alcoholics, because they could not politely refuse the many drinks offered to them. This pressure to entertain visitors and accept their well-meant gifts of vodka and brandy had fatal repercussions: in 1976 Boris hanged himself, almost as if he were allowing Albert to complete his sadistic wartime work, while Alexei's weak health rapidly deteriorated.

Gagarin's wife Valentina successfully raised their two fine daughters, who now enjoy rewarding lives. Valentina still lives within the perimeter of Star City, in a very modest house, and almost never speaks to journalists. Many space veterans regard her humble accommodation as a national disgrace, but she prefers not to draw attention to herself. Golovanov points out, 'She changed very little, despite the lavish attention paid to her by Nikita Khrushchev, who awarded her the Order of Lenin after Gagarin's space flight. Never in her life did she wear it, or any of the awards and medals given to her . . . She was an honest person inside, and so was Gagarin. Despite his fame, he never forgot that he was at the top of a huge pyramid of engineers and constructors who prepared him for his flight.'

This apt metaphor of a pyramid helps illustrate that Gagarin's life was full of contradictions. He was an ambitious and competitive individual, acutely aware that the central achievement of his life was based on the efforts of many others who were not even permitted to reveal their names, let alone share in his public glory. He was a peasant boy at ease with complex engineering equations; a programmed technician who could think for himself; a loyal member of a conformist society who rebelled against the system. He was impetuous, occasionally thoughtless, yet highly disciplined in his work and responsible towards others, often at great risk to himself. He knew little of politics, while displaying a remarkable knack for diplomacy, both at home and abroad. He was an adulterer who never really betrayed his wife and family. As all these conflicting elements of his life intermingle, the story that emerges is one of an essentially decent and brave man giving his best in extraordinary circumstances. He was a hero, in the best and most honest sense of the word.

CHAPTER REFERENCES
AND NOTES

1: Farmboy

1 TASS, *Soviet Man in Space*, Moscow: TASS/Foreign Languages Publishing House, 1961, p. 7. See also: Burchett, Wilfred & Purdy, Anthony, *Cosmonaut Yuri Gagarin*, London: Anthony Gibbs & Phillips, 1961, pp. 87–99.
2 Quoted in Golovanov, Yaroslav, *Our Gagarin*, Moscow: Progress Publishers, 1978, p. 37.
3 Burchett & Purdy, *Cosmonaut Yuri Gagarin*, p. 89.
4 Ibid., p. 90.
5 Golovanov, *Our Gagarin*, p. 42.
6 Burchett & Purdy, *Cosmonaut Yuri Gagarin*, p. 91.
7 Golovanov, *Our Gagarin*, p. 43.
8 Burchett & Purdy, *Cosmonaut Yuri Gagarin*, pp. 92–3
9 Golovanov, *Our Gagarin*, pp. 263–4.
10 Ibid., p. 265.

2: Recruitment

1 Gagarin, *The Road to the Stars*, quoted in Golovanov, *Our Gagarin*, pp. 53–4.
2 Hooper, Gordon R., *The Soviet Cosmonaut Team*, Lowestoft: second edition, GRH Publications, 1990, Vol. II, 'Cosmonaut Biographies', pp. 299–301.
3 Ibid., pp. 161–6.
4 Oberg, James, *Red Star in Orbit*, New York: Random House, 1981, p. 97. Oberg's highly entertaining book was the first popular Western account of the Soviet space programme.
5 Burchett & Purdy, *Cosmonaut Yuri Gagarin*, p. 103.
6 Quoted in Golovanov, *Our Gagarin*, pp. 60–61.
7 Burchett & Purdy, *Cosmonaut Yuri Gagarin*, p. 104.

8 Ibid., p. 104.

3: The Chief Designer

1 For an excellent account of the Chief Designer's life and work, see Harford, James, *Korolev*, New York: John Wiley & Sons, 1997.
2 McCauley, Martin, *Who's Who in Russia Since 1900*, London: Routledge, 1997, pp. 212–13.
3 Burchett & Purdy, *Cosmonaut Yuri Gagarin*, p. 25.
4 Khrushchev, Nikita, *Khrushchev Remembers: The Last Testament*, Boston: Little, Brown, 1970, p. 46.
5 Gatland, Kenneth, *The Illustrated Encyclopedia of Space Technology*, London: second edition, Salamander, 1989. Gatland's book is a reliable general guide to spacecraft designs, launch dates and missions.
6 Interview with Oleg Ivanovsky. The silver cladding around the 'ball' was just a thin layer of reflective foil designed to protect the cabin against harsh solar radiation. The atmospheric heat-shield underneath comprised a much thicker and heavier layer of resin and fibre.
7 Quoted in Golovanov, *Our Gagarin*, p. 89.
8 Quoted in Golovanov, *Our Gagarin*, p. 265.

4: Preparation

1 For a description of Baikonur's construction, see Harvey, Brian, *The New Russian Space Programme*, New York: John Wiley & Sons, 1996, pp. 19–20, Kapustin Yar, pp. 141–3, Plesetsk, pp. 143–40.
2 *Daily Telegraph*, May 6, 1960, p. 1.
3 Secret US space projects are outlined in Trento, Joseph, *Prescription for Disaster*, London: Harrap, 1987, pp. 122–49.
4 Oberg, James, 'Disaster at the Cosmodrome', *Air & Space* Magazine, December 1990, pp. 74–7. For some years Western analysts were confused about the identity of the rocket that blew up. The Nedelin disaster happened to coincide with the early R-7 Mars probe failures, but Nedelin's R-16 was definitely a military missile, not a space launcher. See also: Joint Publications Research Service-UMA-89-015, June 15, 1989, pp. 34–50.
5 Heppenheimer, T.A., *Countdown*, New York: John Wiley & Sons, 1997, pp. 188–9. See also: Harford, *Korolev*, p. 242, and Hooper, *The Soviet Cosmonaut Team*, Vol. 1: 'Background

Sections', pp. 172–3. For a general analysis of Soviet cosmonaut fatalities and hidden rocket failures, see Oberg, James, *Uncovering Soviet Disasters*, London: Robert Hale, 1988, pp. 156–97.

6 Harford, *Korolev*, pp. 163–4.

7 Ibid., pp. 167–8.

8 A full account of the Mercury programme can be found in Swenson, Loyd, Grimwood, James & Alexander, Charles, *This New Ocean: A History of Project Mercury*, Washington, DC: Government Printing Office, NASA SP-4201, 1966.

9 On March 16, 1996, Sotheby's auction house in New York held its second sale of Russian space hardware and memorabilia. Lot 25 consisted of 'Ivan Ivanovich's' capsule, stripped of most of its interior equipment but still clearly recognizable as a Vostok prototype. The catalogue's explanatory notes, prepared with the help of OKB-1 engineer-cosmonaut Konstantin Feoktistov, with additional material by James Oberg and various Russian historical experts, described Ivan's adventures in full. See *Sotheby's Sale Catalogue 6753, Russian Space History*, New York: March 16, 1996.

10 Popescu, Julian, *Russian Space Exploration*, London: Gothard, 1979, p. 16.

5: Pre-Flight

1 Interview with Gai Severin, quoted in Harford, *Korolev*, p. 162.

2 Cameraman Vladimir Suvorov's account of his work can be found in Suvorov, Vladimir & Sabelnikov, Alexander, *The First Manned Spaceflight*, Commack, NY: Nova Science Publishers, 1997, pp. 61–75.

3 Kamanin diaries, April 7, 1961.

4 Suvorov & Sabelnikov, *The First Manned Spaceflight*, p. 58.

5 Harvey, *The New Russian Space Programme*, p. 54.

6 Golovanov, *Our Gagarin*, p. 124.

7 Interview with Gagarin's hairdresser Igor Khoklov.

8 Murray, Charles & Bly Cox, Catherine, *Apollo: The Race to the Moon*, London: Secker & Warburg, 1989, p. 76.

9 Gagarin, *The Road to the Stars*, quoted in Golovanov, *Our Gagarin*, p. 125.

10 Golovanov, *Our Gagarin*, p. 123.

11 Suvorov & Sabelnikov, *The First Manned Spaceflight*, p. 62.

12 Burchett & Purdy, *Cosmonaut Yuri Gagarin*, p. 25.

13 Suvorov & Sabelnikov, *The First Manned Spaceflight*, pp. 64–5.

14 Interview with Oleg Ivanovsky. See also: Hooper, *The Soviet Cosmonaut Team*, Vol. II, pp. 198–9; Oberg, *Uncovering Soviet Disasters*, pp. 157–9.

6: 108 Minutes

1 Quotes taken from original tapes of ground-to-capsule dialogue, quoted in Golovanov, *Our Gagarin*, pp. 127–8, 131–42.
2 A full account of the flight, including extensive quotes from Gagarin himself, can be found in Belyanov, V., *et al.*, 'Tomorrow is Space Programme Day: the Classified Documents on Gagarin's Spaceflight', *Rabochaya Tribuna*, April 11, 1991, pp. 124–8. For an English translation, see Joint Publications Research Service-USP-91–004, September 20, 1991, pp. 71–7. Further details of the flight can be found in 'Yuri Gagarin's Immortal Day', *Spaceflight* magazine, April 1991, pp. 124–8. See also: Baker, David, *The History of Manned Spaceflight*, London: New Cavendish, 1981, pp. 70–73; and Burchett & Purdy, *Cosmonaut Yuri Gagarin*, pp. 110–17 for a highly censored account of the flight.
3 Burchett & Purdy, *Cosmonaut Yuri Gagarin*, p. 143.
4 Golovanov, *Our Gagarin*, pp. 146–7.
5 Ibid., pp. 146–7.
6 Ibid., pp. 149–50.
7 Joint Publications Research Service-USP-91–004, September 20, 1991, pp. 71–7; Belyanov, 'Tomorrow ...' See also: Broad, William J., 'The Untold Perils of the First Manned Spaceflight', *The New York Times*, March 5, 1996.
8 The Vostok flight plan and ejection sequence are described in Newkirk, Dennis, *Almanac of Soviet Manned Spaceflight*, Houston: Gulf, 1990, pp. 7–21.
9 Murray & Cox, *Apollo: The Race to the Moon*, p. 76.
10 Shepard, Alan & Slayton, Deke, *Moonshot*, London: Virgin, 1995, pp. 105–6. See also: *The Times*, April 13, 1961, p. 12, 'We Are Asleep'.
11 Swenson, Grimwood & Alexander, *This New Ocean*, p. 335. See also: 'Ups and downs in Space as US gets set to launch man', *Life* magazine, May 5, 1961.

7: Coming Home

1 Quoted in Golovanov, *Our Gagarin*, p. 150.
2 Golovanov, *Our Gagarin*, pp. 150–51.
3 An English-language report, prepared for the International Aeronautics Federation and signed by Gagarin, Borisenko and other officials, featured as Lot 39 in Sotheby's auction of March 16, 1996. Full details of the document can be found in the reference for Lot 39, *Sotheby's Sale Catalogue 6753, Russian Space History*, March 16, 1996.

4 Golovanov, *Our Gagarin*, p. 151.
5 The conversation between Gagarin and Khrushchev was widely reported in the press. A full transcript appeared in the 1961 TASS pamphlet *Soviet Man in Space*, p. 24.
6 Quoted at length in *The Times* obituary for Gagarin, March 29, 1968.
7 Conversation with the historian Phillip Clark.
8 Gagarin, *The Road to the Stars*, quoted in Golovanov, *Our Gagarin*, pp. 187–8.
9 BBC archive footage, originally from a live broadcast on April 14, 1961 – the first live Western television broadcast from within the USSR – clearly shows the flapping shoelace.
10 See Lynch, Michael, *Stalin and Khrushchev: The USSR, 1924–64*, London: Hodder & Stoughton, 1996, pp. 96–102.
11 Golovanov, *Our Gagarin*, pp. 191–2.
12 Burchett & Purdy, *Cosmonaut Yuri Gagarin*, pp. 118–23.
13 Oberg, *Red Star in Orbit*, p. 55.
14 *Isvestia*, August 28, 1961.

8: The Space Race

1 Murray & Cox, *Apollo: The Race to the Moon*, pp. 77–8.
2 Young, Hugo, *Journey to Tranquillity: The History of Man's Assault on the Moon*, London: Jonathan Cape, 1969, p. 110. In addition, Kennedy's original memo is reproduced as a photograph in the picture section following p. 136.
3 See Trento, *Prescription for Disaster*, pp. 12–13, for a description of Lyndon Johnson's involvement in the creation of NASA in 1958. See also: Lambright, Henry W., *Powering Apollo: James E. Webb of NASA*, Baltimore: Johns Hopkins University Press, 1995, pp. 95–6, 132–5;Archives of Dr John Logsdon, Space Policy Institute, George Washington University, Washington, DC, ref: RG 220, NASC files, Box 17, *Defence 1961, Webb-McNamara Report, 5-8-1961*.
4 Trento, *Prescription for Disaster*, pp. 48–9.
5 Young, *Journey to Tranquillity*, pp. 108–9.
6 Ibid., p. 113.
7 Harford, *Korolev*, p. 178.
8 Ibid. p. 151.
9 Hooper, *The Soviet Cosmonaut Team*, Vol. II, pp. 296–9.

9: The Foros Incident

1 Kamanin's diary entries, September 14–October 3, 1961.
2 Kamanin's diary suggests that Gagarin went out in a motor boat and

'experimented with sharp and dangerous turns'. Anna Rumanseyeva and others remember him going out in a rowing boat, which would explain why he could not easily get back to shore.
3 Kamanin's diaries.
4 A wall chart at Star City commemorates the dates and destinations of all Gagarin's trips. All countries are named, except for the US. Gagarin made a very brief visit to New York on October 16, 1963, but the wall chart refers instead to the 'United Nations'. Gagarin was a guest speaker in the UN complex, and was not formally invited by the US itself.
5 Venyamin Russayev agreed to be interviewed when Valentina Gagarina suggested he should do so. Valentina no longer gives interviews herself.

10: Back to Work

1 Hooper, *The Soviet Cosmonaut Team*, Vol. I, pp. 33–6.
2 Kamanin's diaries, June 22, 1962.
3 Hooper, *The Soviet Cosmonaut Team*, Vol. II, pp. 75–6. See also: Harford, *Korolev*, pp. 165–6.
4 Details of Gagarin's diploma work are derived from extensive interviews with Sergei Belotserkovsky, his tutor at the Zhukovsky Academy.
5 A detailed account of the Voskhod II mission can be found in Harvey, *The New Russian Space Programme*, pp. 82–8. See also: Newkirk, *Almanac of Soviet Manned Spaceflight*, pp. 35–7.
6 Harford, *Korolev*, pp. 49–63.
7 Conversation with James Oberg.

11: Falling to Earth

1 For a valuable account of Mishin's troubles, see Sagdeev, Roald, *The Making of a Soviet Scientist*, New York: John Wiley & Sons, 1994, pp. 123–4, 179–81.
2 The Apollo I fire exposed some scandalous business relationships and incompetencies associated with the NASA moon project. For an eye-opening account, see Young, *Journey to Tranquillity*, pp. 212–48.
3 Archives of Dr John Logsdon, Space Policy Institute, George Washington University, Washington, DC, National Intelligence Estimate Number 11-1-67, March 2, 1967, 'The Soviet Space Programme', p. 18.
4 Quoted in Oberg, *Red Star in Orbit*, pp. 90–91.

5 Yevsikov, Victor, *Re-entry Technology and the Soviet Space Programme: Some Personal Observations*, Reston, VA: Delphic Associates, 1982, quoted in Oberg, *Uncovering Soviet Disasters*, p. 171.
6 Details of Gagarin's involvement in the Soyuz technical assessment are confirmed by ex-KGB officer Venyamin Russayev, at the specific request of Valentina Gagarina.
7 Detailed technical accounts of the possible sequence of failures during Komarov's flight can be found in Newkirk, Dennis, *Almanac of Soviet Manned Spaceflight*, pp. 58–64. See also: Hooper, *The Soviet Cosmonaut Team*, Vol. II, pp. 133–6; Harvey, *The New Russian Space Programme*, pp. 107–10; Gatland, Kenneth, 'The Soviet Space Programme After Soyuz 1', *Spaceflight* magazine, Vol. 9, No. 9, 1967, pp. 298–9; Shepard & Slayton, *Moonshot*, pp. 250–53.
8 The interview, under the headline 'US Electronic Espionage: a Memoir', was published in the left-leaning American journal *Ramparts*, which went out of business in 1980. According to a senior source in the US State Department, the National Security Agency (NSA) considered prosecuting Fellwock. He could have been imprisoned, but the case was dropped because the NSA did not want to admit in open court to their radio monitoring of Soviet space communications.

12: Wreckage

1 Hooper, *The Soviet Cosmonaut Team*, Vol. I, p. 144.
2 Quoted in Golovanov, *Our Gagarin*, p. 214.
3 Ibid., p. 270. Incidentally, this is the only paragraph in a 300-page book to suggest that Gagarin's professional circumstances were sometimes painful for him.
4 Leskov, Sergei, 'The Mystery of Gagarin's Death', *Izvestia*, March 28, 1996.
5 Ibid.
6 Transcripted quotes from original commission reports from the papers of Sergei Belotserkovsky.
7 Ibid.
8 Letter from Igor Kacharovsky, July 3, 1986, from the papers of Sergei Belotserkovsky.
9 The authors are grateful to the Martin Baker company for advice on ejection-seat procedures.
10 Julin, Alexander, 'Gagarin & Serugin – The Last Flight', *Moscow News*, No. 3, week of January 28–February 4, 1996.
11 Interview with Koloshov, *Argumenti i Facti*, No. 12, week of April 2–April 9, 1995.

SELECT BIBLIOGRAPHY

Baker, David, *The History of Manned Spaceflight*, London: New Cavendish, 1981.

Burchett, Wilfred & Purdy, Anthony, *Cosmonaut Yuri Gagarin*, London: Anthony Gibbs & Phillips, 1961.

Gatland, Kenneth, *The Illustrated Encyclopedia of Space Technology*, London: second edition, Salamander, 1989.

Golovanov, Yaroslav, *Our Gagarin*, Moscow: Progress Publishers, 1978.

Harford, James, *Korolev*, New York: John Wiley & Sons, 1997.

Harvey, Brian, *The New Russian Space Programme*, New York: John Wiley & Sons, 1996.

Heppenheimer, T.A., *Countdown*, New York: John Wiley & Sons, 1997.

Hooper, Gordon R., *The Soviet Cosmonaut Team*, Vol. I 'Background Sections', Lowestoft: second edition, GRH Publications, 1990.

Hooper, Gordon R., *The Soviet Cosmonaut Team*, Vol. II 'Cosmonaut Biographies', Lowestoft: second edition, GRH Publications, 1990.

Khrushchev, Nikita, *Khrushchev Remembers: The Last Testament*, Boston: Little, Brown, 1970.

Lambright, Henry W., *Powering Apollo: James E. Webb of NASA*, Baltimore: Johns Hopkins University Press, 1995.

Launius, Roger, *NASA: A History of the US Civil Space Program*, Malabar, Florida: Krieger, 1994.

Lynch, Michael, *Stalin and Khrushchev: The USSR, 1924–64*, London: Hodder & Stoughton, 1996.

McCauley, Martin, *Who's Who in Russia Since 1900*, London: Routledge, 1997.

Murray, Charles & Bly Cox, Catherine, *Apollo: The Race to the Moon*, London: Secker & Warburg, 1989.

Newkirk, Dennis, *Almanac of Soviet Manned Spaceflight*, Houston: Gulf, 1990.

Oberg, James, *Red Star in Orbit*, New York: Random House, 1981.

Oberg, James, *Uncovering Soviet Disasters*, London: Robert Hale, 1988.

Popescu, Julian, *Russian Space Exploration*, London: Gothard, 1979.

Sagdeev, Roald, *The Making of a Soviet Scientist*, New York: John Wiley & Sons, 1994.

Shepard, Alan & Slayton, Deke, *Moonshot*, London: Virgin, 1995.

Sotheby's Sale Catalogue 6753, Russian Space History, New York: March 16, 1996.

Suvorov, Vladimir & Sabelnikov, Alexander, *The First Manned Spaceflight*, Commack, NY: Nova Science Publishers, 1997.

Swenson, Loyd, Grimwood, James & Alexander, Charles, *This New Ocean: a History of Project Mercury*, Washington, DC: GPO, NASA SP-4201, 1966.

TASS, *Soviet Man in Space*, Moscow: TASS/Foreign Languages Publishing House, 1961.

Trento, Joseph, *Prescription for Disaster*, London: Harrap, 1987.

Yevsikov, Victor, *Re-entry Technology and the Soviet Space Programme: Some Personal Observations*, Reston, VA: Delphic Associates, 1982.

Young, Hugo, *Journey to Tranquillity: The History of Man's Assault on the Moon*, London: Jonathan Cape, 1969.

INDEX

INDEX